Graduate Texts in Mathematics
49

T0155951

K. W. Gruenberg

A. J. Weir

Linear Geometry

2nd Edition

Springer-Verlag

New York   Heidelberg   Berlin

K. W. Gruenberg

Department of Pure Mathematics
Queen Mary College
University of London
England

A. J. Weir

School of Mathematical
and Physical Sciences
University of Sussex
England

Editorial Board

P. R. Halmos

Managing Editor
Department of Mathematics
University of California
Santa Barbara, California 93106

F. W. Gehring

Department of Mathematics
University of Michigan
Ann Arbor, Michigan 48104

C. C. Moore

Department of Mathematics
University of California at Berkeley
Berkeley, California 94720

AMS Subject Classification: 50D40

Library of Congress Cataloging in Publication Data

Gruenberg, Karl W
  Linear geometry.

  (Graduate texts in mathematics ; 49)
  1. Geometry, Algebraic. 2. Algebras, Linear.
I. Weir, Alan J., joint author. II. Title.
QA564.G72   1977   516′.35   76–27693

ISBN 978-1-4419-2806-1

© 2010 by Springer-Verlag, New York Inc.

First edition published 1967 by D. Van Nostrand Company.

Printed in the United States of America.

9 8 7 6 5 4 3 2 1

Preface

This is essentially a book on linear algebra. But the approach is somewhat unusual in that we emphasise throughout the geometric aspect of the subject. The material is suitable for a course on linear algebra for mathematics majors at North American Universities in their junior or senior year and at British Universities in their second or third year. However, in view of the structure of undergraduate courses in the United States, it is very possible that, at many institutions, the text may be found more suitable at the beginning graduate level.

The book has two aims: to provide a basic course in linear algebra up to, and including, modules over a principal ideal domain; and to explain in rigorous language the intuitively familiar concepts of euclidean, affine, and projective geometry and the relations between them. It is increasingly recognised that linear algebra should be approached from a geometric point of view. This applies not only to mathematics majors but also to mathematically-oriented natural scientists and engineers.

The material in this book has been taught for many years at Queen Mary College in the University of London and one of us has used portions of it at the University of Michigan and at Cornell University. It can be covered adequately in a full one-year course. But suitable parts can also be used for one-semester courses with either a geometric or a purely algebraic flavor. We shall give below explicit and detailed suggestions on how this can be done (in the "Guide to the Reader").

The first chapter contains in fairly concise form the definition and most elementary properties of a vector space. Chapter 2 then defines affine and projective geometries in terms of vector spaces and establishes explicitly the connexion between these two types of geometry. In Chapter 3, the idea of isomorphism is carried over from vector spaces to affine and projective geometries. In particular, we include a simple proof of the basic theorem of projective geometry, in §3.5. This chapter is also the one in which systems of linear equations make their first appearance (§3.3). They reappear in increasingly sophisticated forms in §§4.5 and 4.6.

Linear algebra proper is continued in Chapter 4 with the usual topics centred on linear mappings. In this chapter the important concept of duality in vector spaces is linked to the idea of dual geometries. In our treatment of bilinear forms in Chapter 5 we take the theory up to, and including, the classification of symmetric forms over the complex and real fields. The geometric significance of bilinear forms in terms of quadrics is taken up in §§5.5–5.7. Chapter 6 presents the elementary facts about euclidean spaces (i.e., real vector spaces with a positive definite symmetric form) and includes the simultaneous reduction theory of a pair of symmetric forms one of which is positive definite (§6.3); as well as the structure of orthogonal transformations (§6.4). The final chapter gives the structure of modules over a polynomial ring (with coefficients in a field) and more generally over a principal ideal domain. This leads naturally to the solution of the similarity problem for complex matrices and the classification of collineations.

We presuppose very little mathematical knowledge at the outset. But the student will find that the style changes to keep pace with his growing mathematical maturity. We certainly do not expect this book to be read in mathematical isolation. In fact, we have found that the material can be taught most successfully if it is allowed to interact with a course on "abstract algebra".

At appropriate places in the text we have inserted remarks pointing the way to further developments. But there are many more places where the teacher himself may lead off in new directions. We mention some examples. §3.6 is an obvious place at which to begin a further study of group theory (and also incidentally, to introduce exact sequences). Chapter 6 leads naturally to elementary topology and infinite-dimensional Hilbert spaces. Our notational use of $\mathscr{A}$ and $\mathscr{P}$ (from Chapter 2 onwards) is properly functorial and students should have their attention drawn to these examples of functors. The definition of projective geometry does not mention partially ordered sets or lattices but these concepts are there in all but name.

We have taken the opportunity of this new edition to include alternative proofs of some basic results (notably in §§5.2, 5.3) and to illustrate many of the main geometric results by means of diagrams. Of course diagrams are most helpful if drawn by the reader, but we hope that the ones given in the text will help to motivate the results and that our hints on the drawing of projective diagrams will encourage the reader to supply his own.

There are over 250 exercises. Very few of these are routine in nature. On the contrary, we have tried to make the exercises shed further light on the subject matter and to carry supplementary information. As a result, they range from the trivial to the very difficult. We have thought it worthwhile to add an appendix containing outline solutions to the more difficult exercises.

We are grateful to all our friends who helped (wittingly and unwittingly) in the writing of this book. Our thanks go also to Paul Halmos for his continuing interest in the book, an interest which has now resulted in the appearance of this new edition.

K. W. Gruenberg
A. J. Weir

April 1977

Contents

Guide to the Reader   ix

Chapter I

Vector Spaces   1

1.1   Sets   1
1.2   Groups, Fields and Vector Spaces   3
1.3   Subspaces   6
1.4   Dimension   8
1.5   The Ground Field   12

Chapter II

Affine and Projective Geometry   15

2.1   Affine Geometries   15
2.2   Affine Propositions of Incidence   17
2.3   Affine Isomorphisms   20
2.4   Homogeneous Vectors   22
2.5   Projective Geometrics   29
2.6   The Embedding of Affine Geometry in
     Projective Geometry   32
2.7   The Fundamental Incidence Theorems of
     Projective Geometry   38

Chapter III

Isomorphisms   42

3.1   Affinities   42
3.2   Projectivities   44
3.3   Linear Equations   49
3.4   Affine and Projective Isomorphisms   52
3.5   Semi-linear Isomorphisms   54
3.6   Groups of Automorphisms   58
3.7   Central Collineations   62

Contents

Chapter IV

Linear Mappings　　　　　　66

4.1　Elementary Properties of Linear Mappings　　66
4.2　Degenerate Affinities and Projectivities　　70
4.3　Matrices　　71
4.4　The Rank of a Linear Mapping　　75
4.5　Linear Equations　　79
4.6　Dual Spaces　　80
4.7　Dualities　　83
4.8　Dual Geometries　　86

Chapter V

Bilinear Forms　　　　　　89

5.1　Elementary Properties of Bilinear Forms　　89
5.2　Orthogonality　　93
5.3　Symmetric and Alternating Bilinear Forms　　97
5.4　Structure Theorems　　102
5.5　Correlations　　110
5.6　Projective Quadrics　　111
5.7　Affine Quadrics　　115
5.8　Sesquilinear Forms　　120

Chapter VI

Euclidean Geometry　　　　　　125

6.1　Distances and Euclidean Geometries　　125
6.2　Similarity Euclidean Geometries　　130
6.3　Euclidean Quadrics　　134
6.4　Euclidean Automorphisms　　140
6.5　Hilbert Spaces　　146

Chapter VII

Modules　　　　　　149

7.1　Rings and Modules　　149
7.2　Submodules and Homomorphisms　　153
7.3　Direct Decompositions　　155
7.4　Equivalence of Matrices over $F[X]$　　163
7.5　Similarity of Matrices over F　　166
7.6　Classification of Collineations　　170

Solutions　　　　　　175
List of Symbols　　　　　　188
Bibliography　　　　　　191
Index　　　　　　193

Guide to the Reader

This book can be used for linear algebra courses involving varying amounts of geometry. Apart from the obvious use of the whole book as a one year course on algebra and linear geometry, here are some other suggestions.

(A)   *One semester course in basic linear algebra*

All of Chapter I.

§2.1 for the definition of coset and dimension of coset.

§3.3.

Chapter IV, but omitting §§4.2, 4.7, 4.8.

(B)   *One semester course in linear geometry*

Prerequisite: (A) above or any standard introduction to linear algebra.

All of Chapter II.

§§3.1–3.4.

Then either (B$_1$) or (B$_2$) (or, of course, both if time allows):

(B$_1$)   §§4.7, 4.8.

Chapter V up to the end of Proposition 12 (p. 118).

(B$_2$)   §§5.1–5.4 (see Note 1, below; also Notes 2, 3, for saving time); §5.6 but omitting Proposition 10; §5.7 to the end of Proposition 12. §§6.1, 6.3, 6.4 to the "orientation" paragraph on p. 144.

(C)   *One year course in linear algebra*

The material in (A) above.

Chapter V but omitting §§5.5–5.7.

Chapter VI but *omitting* the following: §6.1 from mid p. 127 (where distance on a coset is defined); §6.2; §6.3 from Theorem 2 onwards;

§6.4 from mid p. 144 (where similarity classes of distances are defined).

Chapter VII, but omitting §7.6.

Notes

1. §§5.1–5.3 can be read without reference to dual spaces. In particular, the first proof of Proposition 4 (p. 93) and the first proof of Lemma 4 (p. 97) are then the ones to read.

2. The reader who is only interested in symmetric or skew-symmetric bilinear forms can ignore the distinction between $\perp$ and $\top$. This will simplify parts of §5.2, and in §5.3 the notion of orthosymmetry and Proposition 6 can then be omitted.

3. In §5.3, the question of characteristic 2 arises naturally when skew-symmetric forms are discussed. But the reader who wishes to assume $2 \neq 0$ in the fields he is using can omit the latter part of §5.3.

CORRECTION

Graduate Texts in Mathematics, Vol. 49
K. W. Gruenberg and A. J. Weir: *Linear Geometry*

Page 62, last two lines should read:

If H and K are two distinct hyperplanes of invariant points for a collineation π, then $\pi = \mathscr{P}(f)$ where the restrictions of f to H and K are

CHAPTER I

Vector Spaces

There are at least two methods of defining the basic notions of geometry. The one which appears more natural at first sight, and which is in many ways more satisfactory from the logical point of view, is the so called *synthetic* approach. This begins by postulating objects such as points, lines and planes and builds up the whole system of geometry from certain *axioms* relating these objects. In order to progress beyond a few trivial theorems, however, there must be sufficient axioms. Unfortunately, it is difficult to foresee the kind of axioms which are required in order to be able to prove what one regards as "fundamental" theorems (such as Pappus' Theorem and Desargues' Theorem). This method is very difficult an introduction to the subject.

The second approach is to base the geometry on an *algebraic* foundation. We favor this approach since it allows us to "build in" enough axioms about our geometry at the outset. The axioms of the synthetic geometry now become theorems in our algebraic geometry. Moreover, the interdependence of the algebraic and geometric ideas will be seen to enrich both disciplines and to throw light on them both. (For an introduction to the synthetic approach the reader may consult [4], [9].)

1.1 Sets

We shall not define the basic notion of *set* (or collection, or aggregate) which we regard as intuitive. Further, we shall assume that the reader is familiar with the simplest properties of the set $\mathbf{Z}$ of integers (positive, negative and zero), the set $\mathbf{Q}$ of rational numbers, the set $\mathbf{R}$ of real numbers, the set $\mathbf{C}$ of complex numbers and the set $\mathbf{F}_p$ of integers modulo a prime p.

As a shorthand for the statement "x is an element of the set S" we shall write $x \in S$. If S and T are two sets with the property that every element of S is an element of T, we write $S \subset T$ and say "S is contained in T" or that "S is a subset of T"; equivalently, we also write $T \supset S$ and say "T contains S". Note that according to this definition $S \subset S$. We shall say that the sets S and T are *equal*, and write $S = T$, if $S \subset T$ and $S \supset T$. If the set S consists of the elements $x, y, \ldots$ then we write $S = \{x, y, \ldots\}$. Thus, for example, $x \in S$ if, and only if, $\{x\} \subset S$. Strokes through symbols usually give the negative: for example, $\neq$, $\not\subset$, $\notin$ stand for, respectively, "is not equal to", "is not contained in", "is not an element of".

If S and T are given sets then a *mapping* (or *function*) f of S into T is a rule which associates to each element s in S a unique element sf in T. In these circumstances we shall often write $f: S \to T$ or $f: s \to sf$. The element sf is called the *image* of the element s under f. (It is often also written as $f(s)$ or s^f or f_s, whichever is the most convenient notation for the purpose at hand.) The set of all sf as s varies in S is the *image* of S under f, or simply the *image of f*, and is denoted by Sf. If $Sf = T$ then we say that f is a mapping of S *onto* T; if the images under f of any two distinct elements of S are distinct elements of T, then f is a *one–one* mapping. (The above definition of mapping seems to involve the undefined notion of "rule". A more sophisticated definition can be given in terms of the "graph" of f which is the set of all ordered pairs (s, sf) for s in S. The definition can thus be thrown back on the basic concept of set.)

The notation sf for the image is particularly suitable if mappings are to be combined; more specifically, if f is a mapping of S into T and if g is a mapping of T into U, then the *product fg* is the mapping of S into U defined by the equation $s(fg) = (sf)g$. In other words fg stands for "apply first f, then g". If one uses the functional notation, then $(fg)(s) = g(f(s))$.

On the other hand, there is a situation in which the *index notation f_s* for the image is better than sf. This occurs when we are interested in "listing" the elements of the image of S under f. It is then usual to refer to S as the *set of indices*, to call f_s the *s-term* of f, to write f in the alternative form $(f_s)_{s \in S}$ and to call f a *family* rather than a mapping. As s runs through the set S of indices the terms f_s run through the image of S under f. It is important to realize that some of the terms will be repeated unless the mapping f is one–one.

If an arbitrary non-empty set S is given then we can always "list" the elements of S by using the identity mapping 1_S, which sends each element of S into itself.

We mention two familiar examples of this notation:

1. If I is the set of all positive integers, then $(f_i)_{i \in I}$ is a *sequence*.
2. If I is the finite set $\{1, 2, \ldots, n\}$, then $(f_i)_{i \in I}$ is an *n–tuple*.

If we take note of the natural ordering of the integers, then the sequences and n-tuples above are called *ordered sequences* and *ordered n-tuples* respectively.

DEFINITION. Suppose that $(M_i)_{i \in I}$ is a family where each M_i is itself a set. We define the *intersection* $\cap(M_i : i \in I)$ of this family to be the set of all elements which belong to every M_i $(i \in I)$; also, the *union* $\cup(M_i : i \in I)$ is the set of all elements which belong to at least one of the sets M_i $(i \in I)$. Observe that, for this definition to make sense, I cannot be empty.

The intersection and union of the sets $M_1, \ldots, M_n$ are usually denoted by $M_1 \cap \cdots \cap M_n$ and $M_1 \cup \cdots \cup M_n$, respectively.

If we are simply given a (non-empty) set S whose elements M are sets, then we index S by means of the identity mapping and use the above definition. The intersection $\cap(M : M \in S)$ is therefore the set of all elements which belong to every set M in S and the union $\cup(M : M \in S)$ is the set of all elements which belong to at least one set M in S.

EXERCISES

1. Let S, T be sets, f a mapping of S into T, g a mapping of T into S, and denote by 1_S, 1_T, respectively, the identity mappings on S and T. Prove that (i) $fg = 1_S$ implies that f is one–one; and (ii) $gf = 1_T$ implies that f is onto T. Show that if (i) and (ii) hold, then g is uniquely determined by f. (In this case we write $g = f^{-1}$ and call g the *inverse* of f.)

2. The population on an island is greater than the number of hairs on the head of any one inhabitant. Show that if nobody is bald then at least two people have the same number of hairs on their heads.

3. If f, g, h are mappings of S into T, T into U, U into V, respectively, show that $(fg)h = f(gh)$.

1.2 Groups, Fields and Vector Spaces

We assume that the reader is to some extent familiar with the concept of a vector in three dimensional euclidean space. He will know that any two vectors can be added to give another vector and any vector can be multiplied by a real number (or scalar) to give yet another vector. The primary object of this book is to generalize these ideas and to study many of their geometrical properties. In order to axiomatize the addition of vectors we introduce the concept of a *group*, and in order to generalize the notion of scalar we define a general *field*. The reader who is meeting these abstract ideas for the first time may find it helpful at the beginning to replace the general field F in our definition of a vector space by the familiar field **R** of real numbers and to accept the other axioms as a minimum set of sensible rules which allow the usual manipulations with vectors and scalars. The subsequent sections have more geometric and algebraic motivation and are not likely to cause the same difficulty.

DEFINITION. Let G be a set together with a rule (called *multiplication*) which associates to any two elements a, b in G a further element ab in G (called the *product* of a and b). If the following axioms are satisfied then G is called a *group*:

G.1. $(ab)c = a(bc)$ for all a, b, c in G;

G.2. there exists a unique element 1 in G (called the *identity*) such that $a1 = 1a = a$ for all a in G;

G.3. for each element a in G there exists a unique element a^{-1} in G (called the *inverse* of a) such that $a\,a^{-1} = a^{-1}a = 1$.

A subset of G which is itself a group (with respect to the same rule of multiplication as G) is called a *subgroup* of G. Note that, in particular, G is a subgroup of G and so also is $\{1\}$, called the *trivial subgroup* of G.

We shall see that the order of the elements a, b in the product ab is important (see exercises 2, 3 below). This leads to a further definition.

DEFINITION. A group G is *commutative* (or *abelian*) if

G.4. $ab = ba$ for all a, b in G.

It is only a matter of convenience to make use of the words "multiplication", "product", "inverse" and "identity" in the definition of a group. Sometimes other terminologies and notations are more useful. The most important alternative is to call the given rule "addition", and to replace the product ab by the *sum* $a + b$, the identity 1 by the *zero* 0 and the inverse a^{-1} by the *negative* $-a$ (minus a). When this notation and terminology is used we shall speak of G as a *group with respect to addition*; while if that of our original definition is employed we shall say that G is a *group with respect to multiplication*.

The following are important examples of groups:

1. The integers $\mathbf{Z}$ form a commutative group with respect to addition. The set $\mathbf{Z}^*$ of non-zero integers, however, is not a group with respect to multiplication.

2. The sets $\mathbf{Q}$, $\mathbf{R}$, $\mathbf{C}$, $\mathbf{F}_p$ are commutative groups with respect to addition. The set of non-zero elements in each of these sets is a commutative group with respect to multiplication.

3. If S is the set of all points in three dimensional euclidean space, then the rotations about the lines through a fixed point O of S may be regarded as mappings of S into S. These rotations form a group with respect to (mapping) multiplication. (The rigorous definitions of these concepts will be given in Chapter VI.)

4. The set of all *permutations* of a set S (i.e., all one–one mappings of S onto S) is a group with respect to (mapping) multiplication.

DEFINITION. Let F be a set together with rules of addition and multiplication which associate to any two elements x, y in F a sum $x + y$

and a product xy, both in F. Then F is called a *field* if the following axioms are satisfied:

F.1. F is a commutative group with respect to addition;

F.2. the set F^*, obtained from F by omitting the zero, is a commutative group with respect to multiplication;

F.3. $x(y+z) = xy + xz$ and $(y+z)x = yx + zx$ for all x, y, z in F.

The fields that we shall mainly have in mind in this book are the fields $\mathbf{Q}$, $\mathbf{R}$, $\mathbf{C}$, $\mathbf{F}_p$, listed in example 2 above.

We are now ready to define the basic object of our study.

DEFINITION. Let F be a given field and V a set together with rules of addition and multiplication which associate to any two elements a, b in V a *sum* $a+b$ in V, and to any two elements x in F, a in V a *product* xa in V. Then V is called a *vector space over the field* F if the following axioms hold:

V.1. V is a commutative group with respect to addition;

V.2. $x(a+b) = xa + xb$,

V.3. $(x+y)a = xa + ya$,

V.4. $(xy)a = x(ya)$,

V.5. $1a = a$ where 1 is the identity element of F,

for all x, y in F and all a, b in V.

We shall refer to the elements of V as *vectors* and the elements of F as *scalars*. The only notational distinction we shall make between vectors and scalars is to denote the zero elements of V and F by 0_V and 0_F respectively. Since $x0_V = 0_V$ for all x in F and $0_F a = 0_V$ for all a in V (see exercise 7 below) even this distinction will almost always be dropped and 0_V, 0_F be written simply as 0.

The following examples show how ubiquitous vector spaces are in mathematics. The first example is particularly important for our purposes in this book.

1. Let F be a field and denote by F^n the set of all n-tuples $(x_1, \ldots, x_n)$ where $x_1, \ldots, x_n \in F$. We define the following rules of addition and multiplication:

$$(x_1, \cdots, x_n) + (y_1, \cdots, y_n) = (x_1 + y_1, \cdots, x_n + y_n),$$

$$x(x_1, \cdots, x_n) = (xx_1, \cdots, xx_n)$$

for all x, $x_1, \ldots, x_n$, $y_1, \ldots, y_n$ in F.

With these rules F^n is a vector space over F.

2. If F is a subfield of a field E, then E can be regarded as a vector space over F in the following way: E is already a group with respect to addition and we define the product of an element a of E (a "vector")

by an element x of F (a "scalar") to be xa, their ordinary product as elements of E. It is important to note that the elements in F now have two quite distinct parts to play: on the one hand, as elements of F, they are scalars; but on the other hand, as elements of the containing field E, they are vectors.

We mention some special cases: F is always a vector space over F; $\mathbf{R}$ is a vector space over $\mathbf{Q}$; $\mathbf{C}$ is a vector space over $\mathbf{R}$ and also over $\mathbf{Q}$.

3. The set $F[X]$ of all polynomials in the indeterminate X with coefficients in the field F is a vector space over F.

4. The set of all Cauchy sequences with elements in $\mathbf{Q}$ is a vector space over $\mathbf{Q}$.

EXERCISES

1. Extend rule G.1 to show that parentheses are unnecessary in a product (or sum) of any finite number of elements of a group.

2. Show that the rotation group of example 3 is not commutative.

3. Show that the permutation groups of example 4 are not commutative if S contains more than two elements.

4. Show from the axioms that a field must contain at least two elements. Write out the addition and multiplication tables for a field with just two elements.

5. Let V be a vector space over the field F. Show that any finite *linear combination* $x_1a_1 + x_2a_2 + \cdots + x_ra_r = \sum_{i=1}^{r} x_ia_i$, where the x_i's are scalars and the a_i's are vectors, can be written unambiguously without the use of parentheses.

6. If $a, b \in V$ show that the equation $v + b = a$ has a unique solution v in V. This solution is denoted by $a - b$.

7. Show that $0_Fa = 0_V$ for all a in V and $x0_V = 0_V$ for all x in F. Show further that $(-1)a = -a$ for every a in V. (For the first equation use $(0_F + 0_F)a = 0_Fa$ and exercise 6.)

1.3 Subspaces

DEFINITION. A subset M of a vector space V over F is called a *subspace* of V if M is a vector space over F in its own right, but with respect to the same addition and scalar multiplication as V.

Observe that a vector space contains, by definition, a zero vector and so a subspace can never be empty. In order to check that a (nonempty) subset M is a subspace, we need only verify that if $a, b \in M$ and $x \in F$, then $a + b \in M$ and $xa \in M$. (In verifying that M is a subgroup of V with respect to addition we use the fact that $(-1)a = -a$.)

Our definition clearly ensures that V is a subspace (of itself). At the other extreme, the set $\{0_V\}$ is a subspace, called the *zero subspace*, and we usually write this simply as 0_V. It is also clear that 0_V is contained in every subspace of V.

We shall base our construction of subspaces on the following simple proposition.

PROPOSITION 1. *The intersection of any family of subspaces of V is again a subspace of V.*

PROOF. Put $M = \bigcap(M_i : i \in I)$. Since $0_V \in M_i$ for each i, M is not empty. If $a, b \in M$ and $x \in F$, then $a + b \in M_i$ and $xa \in M_i$ for each i; thus $a + b \in M$ and $xa \in M$.

DEFINITION. If S is any set of vectors in V then we denote by $[S]$ the intersection of all the subspaces of V which contain S. (There is always at least one subspace containing S, namely V itself, and so this intersection is defined.) The subspace $[S]$ is the "least" subspace containing S, in the sense that if K is any subspace containing S, then K contains $[S]$. We say that $[S]$ is the subspace *spanned* (or *generated*) by S.

This definition of $[S]$ may be referred to as the definition "from above". Provided that S is not empty there is an equivalent definition "from below": We form the set M of all (finite) linear combinations $x_1 s_1 + \cdots + x_r s_r$, where $x_i \in F$ and $s_i \in S$. It is immediately seen that M is a subspace of V; that M contains S; and, in fact, that any subspace K of V which contains S must contain all of M. Hence $M = [S]$.

It might be expected that the union of several subspaces of V would be a subspace, but it is easily shown by means of examples that this is not the case. We are thus led to the following definition.

DEFINITION. The *sum* (or *join*) of any family of subspaces is the subspace spanned by their union. In other words, the sum of the subspaces M_i ($i \in I$) is the least subspace containing them all. The sum is denoted by $+(M_i : i \in I)$ or simply $+(M_i)$. The sum of subspaces $M_1, \ldots, M_n$ is usually written $M_1 + M_2 + \cdots + M_n$.

DEFINITION. If $M \cap N = 0$ we call $M + N$ the *direct sum* of M and N, and write it as $M \oplus N$. More generally, the sum of the subspaces M_i ($i \in I$) is *direct*, and written $\bigoplus(M_i : i \in I)$ if, for each j in I, we have $+(M_i : i \in I, i \neq j) \cap M_j = 0$.

Note that the condition above on the subspaces M_i is stronger than the mere requirement that $M_i \cap M_j = 0$ for all $i \neq j$ (see exercise 7). If $v \in M_1 \oplus \cdots \oplus M_r$, then v can be written *uniquely* in the form $v = m_1 + \cdots + m_r$, where $m_i \in M_i$ for $i = 1, \ldots, r$.

EXERCISES

1. If M is a subspace of V, show that the zero vector of M is the same as the zero vector of V.

2. Show that the set of all polynomials $f(X)$ in $F[X]$ satisfying $f(x_i)=0$ $(i=1,\ldots,r)$ for given $x_1,\ldots,x_r$ in F is a subspace of $F[X]$. Show that the set of all polynomials of degree less than n (including the zero polynomial) is a subspace of $F[X]$.

3. If the mn elements a_{ij} $(i=1,\ldots,m;j=1,\ldots,n)$ lie in a field F, show that the set of all solutions $(x_1,\ldots,x_n)$ in F^n of the linear equations

$$\sum_{j=1}^{n} a_{ij}X_j = 0 \quad (i = 1,\ldots,m)$$

   is a subspace of F^n.

4. Give precise meaning to the statement that $M \cap N$ is the "greatest" subspace of V contained in both M and N.

5. Give an example of two subspaces of a vector space whose union is not a subspace.

6. Prove that $v \in M + N$ if, and only if, v can be expressed in the form $m+n$ where $m \in M$, $n \in N$; and

$$v \in +(M_i : i \in I) \quad \text{if, and only if,} \quad v = m_{i_1} + \cdots + m_{i_r}$$

   for some i_k in I and m_{i_k} in M_{i_k} $(k=1,\ldots,r)$.

7. Consider the following subsets of F^3: M_1 consists of all $(x,0,0)$; M_2 consists of all $(0,x,0)$; and M_3 consists of all (x,x,y), where $x,y \in F$. Show that M_1, M_2, M_3 are subspaces of F^3 which satisfy $M_i \cap M_j = 0$ for all $i \neq j$; that $F^3 = M_1 + M_2 + M_3$; but that F^3 is not the direct sum of M_1, M_2, M_3.

8. Prove the last statement made before these exercises.

9. If V is a vector space over $\mathbf{F}_p$, prove that every subgroup of V with respect to addition is a subspace.

1.4   Dimension

The essentially geometrical character of a vector space will emerge as we proceed. Our immediate task is to define the fundamental concept of dimension; and we shall do this by using the even more primitive notion of *linear dependence*.

DEFINITION. If S is a subset of a vector space V over a field F, then we say that the vectors of S are *linearly dependent* if the zero vector of V is a non-trivial linear combination of distinct vectors in S; i.e., if there exist scalars $x_1,\ldots,x_r$ in F, not all 0, and distinct vectors $s_1,\ldots,s_r$ in S such that $x_1s_1 + \cdots + x_rs_r = 0$. (In these circumstances we shall sometimes say that the set S is linearly dependent.) If the vectors of S are not linearly dependent then they are *linearly independent*. The vector v is said to be *linearly dependent on* the vectors of S if v is a (finite) linear combination of vectors in S.

The connexions between these three concepts are given in the following proposition.

PROPOSITION 2. (1) *If $a_1, \ldots, a_r$ where $r \geq 2$ are linearly dependent, then one of the a_i is linearly dependent on the rest.*

(2) *If $b_1, \ldots, b_s$ are linearly independent, but $v, b_1, \ldots, b_s$ are linearly dependent, then v is linearly dependent on $b_1, \ldots, b_s$.*

PROOF. (1) We are given that $x_1 a_1 + \cdots + x_r a_r = 0$ for some $x_1, \ldots, x_r$ in F, not all zero. Suppose that $x_j \neq 0$; then we may multiply by x_j^{-1} and express a_j as a linear combination of the remaining a_i's.

(2) We are given $xv + y_1 b_1 + \cdots + y_s b_s = 0$ for some $x, y_1, \ldots, y_s$ in F, not all zero. Since $b_1, \ldots, b_s$ are linearly independent, x cannot be zero. We multiply by x^{-1} and express v as a linear combination of the b_i's.

DEFINITION. A *basis* (or *base*) of a vector space V is a set of linearly independent vectors in V which also span V.

DEFINITION. A vector space which can be spanned by a finite number of vectors is called *finite dimensional*; otherwise it is said to be *infinite dimensional*.

We shall confine our attention in almost all of this book to finite dimensional vector spaces and, in fact, from Chapter II onwards we make a convention to this effect. The reader will see, however, that many of the results (and most of the definitions) do not need this restriction.

PROPOSITION 3. *Any finite dimensional vector space contains a basis.*

PROOF. Suppose that $a_1, \ldots, a_r$ span V. If these vectors are linearly dependent and $r \geq 2$, then by Proposition 2 (1) one of these a_i is linearly dependent on the rest. If this vector is discarded then the rest still span V. Continuing in this way we either arrive at a basis or at a single vector a, where a is linearly dependent. This means that $a = 0$. Hence $V = 0$ and V is spanned by the empty set.

All the remaining results in this section depend directly on the following fundamental result.

EXCHANGE LEMMA. *If $a_1, \ldots, a_r$ span V and if $b_1, \ldots, b_s$ are linearly independent in V, then $s \leq r$ and, by renumbering the a_i's if necessary, $b_1, \ldots, b_s, a_{s+1}, \ldots, a_r$ span V.*

PROOF. We use induction on k to prove that $b_1, \ldots, b_k, a_{k+1}, \ldots, a_r$ span V (so long as $k \leq s$). When $k = 0$ this is given (in the sense that no b_i's appear). Assume the statement for $k = h - 1$ and consider the case $k = h$. Then $b_h = y_1 b_1 + \cdots + y_{h-1} b_{h-1} + x_h a_h + \cdots + x_r a_r$, and the linear independence of the b_i's ensures that at least one x_i is non-zero. By renumbering the vectors $a_h, \ldots, a_r$, if necessary, we may suppose that $x_h \neq 0$. Now we can solve for a_h as an element of

$$[b_1, \ldots, b_h, a_{h+1}, \ldots, a_r] = M,$$

say, where M contains $b_1, \ldots, b_{h-1}, a_{h+1}, \ldots, a_r$ and also a_h. Hence $M = V$, by the induction hypothesis. The result now follows by induction.

THEOREM 1. *Let V be a given finite dimensional vector space.*
(1) *Any two bases of V have the same number of elements.*
(2) *The following statements concerning the finite subset S of V are equivalent:*
 (i) *S is a basis of V,*
 (ii) *S is a minimal spanning set of V,*
 (iii) *S is a maximal linearly independent set in V.*
(3) *Any set of linearly independent vectors in V can be extended to give a basis of V.*

PROOF. (1) By the Exchange Lemma, if $\{a_1, \ldots, a_r\}$ and $\{b_1, \ldots, b_s\}$ are bases of V, then $s \leq r$ and $r \leq s$.
(2) If $\{b_1, \ldots, b_s\}$ is a basis, then the inequality $s \leq r$ of the Exchange Lemma shows that this basis is a minimal spanning set. Thus (i) implies (ii). If $\{a_1, \ldots, a_r\}$ is a basis, then this same inequality shows that this basis is a maximal linearly independent set. Thus (i) implies (iii). If S is a minimal spanning set, then by the proof of Proposition 3, S must be linearly independent. This shows that (ii) implies (i). Finally, if S is a maximal linearly independent set, then by Proposition 2 (2), S is a spanning set, and so (iii) implies (i).
(3) Suppose that $\{a_1, \ldots, a_r\}$ is a basis and $b_1, \ldots, b_s$ are the given linearly independent vectors. Then by the Exchange Lemma, renumbering if necessary, $b_1, \ldots, b_s, a_{s+1}, \ldots, a_r$ span V. But this spanning set contains r vectors and so is a basis by part (2) of the theorem.

We are now in a position to give the promised definition of dimension.

DEFINITION. Let V be a finite dimensional vector space over the field F. The number of elements in a basis of V is the *dimension* of V (over F) and is written $\dim_F V$, or simply $\dim V$ if the field of scalars F is understood.

We have the following immediate corollaries of Theorem 1.

COROLLARY 1. *If K is a subspace of V, then any basis of K can be extended to a basis of V.*

COROLLARY 2. *If K is a subspace of V then there always exists at least one subspace M such that*

$$M \oplus K = V.$$

The following result will have important applications in the geometry which we introduce in the next chapter.

THEOREM 2. *Let M and N be finite dimensional subspaces of a vector space.*
(1) *If $M \subset N$, then $\dim M \leq \dim N$ and $\dim M = \dim N$ implies $M = N$.*
(2) *Both $M + N$ and $M \cap N$ are finite dimensional and*

$$\dim (M + N) + \dim (M \cap N) = \dim M + \dim N.$$

PROOF. Part (1) is an immediate consequence of Theorem 1. To prove part (2), we may suppose, using Theorem 1, Corollary 1, that $\{a_1, \ldots, a_r\}$ is a basis of $M \cap N$, $\{a_1, \ldots, a_r, b_1, \ldots, b_s\}$ is a basis of M and that $\{a_1, \ldots, a_r, c_1, \ldots, c_t\}$ is a basis of N. It is clear that $M + N$ is spanned by $a_1, \ldots, a_r, b_1, \ldots, b_s, c_1, \ldots, c_t$ and it only remains to prove that these vectors are linearly independent.

Suppose that

$$x_1 a_1 + \cdots + x_r a_r + y_1 b_1 + \cdots + y_s b_s + z_1 c_1 + \cdots + z_t c_t = 0,$$

where the x_i's, y_i's and z_i's are in F. Thus, in an obvious shorthand, we suppose that $a + b + c = 0$. Then $c = -(a + b)$ is an element of $M \cap N$. But the a_i's and c_i's are linearly independent and so the z_i's are zero. Now the a_i's and b_i's are linearly independent and so all the x_i's and y_i's are zero.

EXERCISES

1. If a set S of vectors contains 0 then S is linearly dependent.

2. If a set S of vectors is linearly independent then so is every subset of S (including, by convention, the empty set).

3. The vectors $v_1, v_2, \ldots$ are linearly independent if, and only if, $v_1, v_2, \ldots, v_k$ are linearly independent for all $k > 0$.

4. If S_1, S_2, S_3 are three sets of vectors such that every element of S_1 is linearly dependent on S_2 and every element of S_2 is linearly dependent on S_3, then every element of S_1 is linearly dependent on S_3.

5. Check the sets $\{1 - X, X(1 - X), 1 - X^2\}$, $\{1, X, X^2, \ldots\}$ for linear dependence in $F[X]$.

6. Prove that the vectors $(1, 0, 0)$, $(0, 1, 0)$, $(0, 0, 1)$, $(1, 1, 1)$ are linearly dependent in F^3 but that any three of them are linearly independent.

7. Show that 1, i form a basis of $\mathbf{C}$ over $\mathbf{R}$ (where $i^2 = -1$). Prove also that $\mathbf{R}$ is infinite dimensional over $\mathbf{Q}$. (Hint: A finite dimensional vector space over $\mathbf{Q}$ is enumerable.)

8. Show that 1, X, $X^2, \ldots$ form a basis of $F[X]$ over F.

9. Let e_i be the n-tuple whose i-term is 1 and whose other terms are all 0 $(i = 1, \ldots, n)$. Show that $e_1, \ldots, e_n$ form a basis of F^n. We call this the *standard basis* of F^n.

10. Prove that a vector space V is finite dimensional if, and only if, there is a finite maximum for the lengths k of all possible chains of subspaces

$$M_0 \supset M_1 \supset \cdots \supset M_k$$

where $M_{i-1} \neq M_i$ for each $i = 1, \ldots, k$. Show further that if there is such a maximum, then it is precisely dim V.

11. If $\{a_1, \ldots, a_n\}$ is a basis of V and if A_i is the one-dimensional subspace spanned by a_i $(i = 1, \ldots, n)$, then $V = A_1 \oplus \cdots \oplus A_n$.

12. Find an example which shows that the subspace M of Theorem 1, Corollary 2, is not necessarily unique.

13. If V is a finite dimensional vector space, prove that the sum $M + N$ is direct if, and only if, dim $(M + N) = \dim M + \dim N$. More generally, show that the sum $M_1 + \cdots + M_r$ is direct if, and only if, we have dim $(M_1 + \cdots + M_r) = \dim M_1 + \cdots + \dim M_r$.

1.5 The Ground Field

If V is a vector space over a field F, then we often refer to F as the *ground field*. It is clear that the definition of a vector space depends on the particular field chosen as ground field. In the case where V has finite dimension n over F we shall show that, as an abstract vector space, V is effectively the same as the vector space F^n. To be more precise we need a definition.

DEFINITION. Let V and V' be two vector spaces over the same field F. A one–one mapping f of V onto V' which preserves addition and multiplication by scalars is called an *isomorphism of V onto V'*. If such an isomorphism exists then we say that V and V' are *isomorphic* or that V is *isomorphic to V'*.

The mapping f satisfies the rules:
(1) $(a + b)f = af + bf$,
(2) $(xa)f = x(af)$
for all a, b in V and all x in F.

THEOREM 3. *If V and V' are vector spaces over F of finite dimension n, then V and V' are isomorphic. In particular, V and F^n are isomorphic.*

PROOF. Let $\{a_1, \ldots, a_n\}$ and $\{a_1', \ldots, a_n'\}$ be bases over F for V and V', respectively. Then each v in V can be expressed uniquely in the form

$$v = x_1 a_1 + \cdots + x_n a_n$$

where $x_1, \ldots, x_n \in F$. The mapping f which takes

$$x_1 a_1 + \cdots + x_n a_n \text{ in } V \text{ to } x_1 a_1' + \cdots + x_n a_n' \text{ in } V'$$

satisfies conditions (1) and (2) above. It is clearly one–one and onto V' and so is an isomorphism of V onto V'.

We remark that the isomorphism constructed in the proof of Theorem 3 is by no means unique as it depends on our choice of bases for V and V'.

An important special property of the vector space F^n, not shared by general vector spaces, is that F^n has a built-in natural basis, the *standard basis* $\{e_1, \ldots, e_n\}$, where e_i is the n-tuple whose i-term is 1 and whose other terms are all 0 (cf. exercise 9 of the previous section).

When an ordered basis $(a_1, \ldots, a_n)$ of V is given there is a unique isomorphism f of V onto F^n such that $a_i f = e_i$ for $i = 1, \ldots, n$, viz., the mapping

$$x_1 a_1 + \cdots + x_n a_n \to (x_1, \ldots, x_n).$$

Conversely, if an isomorphism f of V onto F^n is given, then there is a unique ordered basis $(a_1, \ldots, a_n)$ of V such that $a_i f = e_i$ for all i. The mapping f is a rule which associates to any vector $v = x_1 a_1 + \cdots + x_n a_n$ a *coordinate row* $(x_1, \ldots, x_n)$ and so we call f a *coordinate system* of V. It is clear that there is a one–one correspondence between coordinate systems f of V and ordered bases $(a_1, \ldots, a_n)$ of V defined by the rule

$$f\colon x_1 a_1 + \cdots + x_n a_n \to (x_1, \ldots, x_n).$$

Theorem 3 deals with the relationship between finite dimensional vector spaces over the same field. We must now mention briefly a subtler question: what happens to a vector space if we restrict or extend the ground field?

If the field E contains the field F then any vector space V over E can be regarded as a vector space over F, simply by ignoring the scalars in E which are not in F. Let us, for the moment, distinguish these vector spaces by writing them as V_E, V_F respectively. It is important to note that, although V_E and V_F consist of exactly the same vectors, their dimensions will usually be different. For example, if V_C is a vector space of dimension n over $\mathbf{C}$, then the vector space $V_\mathbf{R}$, obtained by restricting the ground field to $\mathbf{R}$ in this way, has dimension $2n$.

There is another method of restricting the ground field. If V_E has dimension n over E then we may pick a basis $\{a_1, \ldots, a_n\}$ of V_E. The set of all vectors $x_1a_1 + \cdots + x_na_n$ where the x_i's belong to F is a vector space U_F over F of the same dimension n. This second method is in fact more useful from our point of view, but it has the severe disadvantage of not being unique. In particular, if V_C is a non-zero vector space over C, we shall not expect to be able to find a unique subset U_R of "real vectors" in V_C. The fact that we *can* do so for the vector space C^n itself is due to the circumstance that C^n has a standard basis.

In our study of geometry we shall frequently be interested in vector spaces over R and C. Let U be a given vector space of dimension n over R. We now show how to construct, *in a unique way*, a vector space $U_{(C)}$ over C, also of dimension n, which contains an isomorphic copy of U.

We recall that C is the set of all ordered pairs (x, y) where $x, y \in R$, subject to the rules

$$(x, y) + (x', y') = (x + x', y + y'),$$

$$(x, y)(x', y') = (xx' - yy', xy' + yx'),$$

and that R is identified with the set of all ordered pairs $(x, 0)$. We may then write $(0, 1) = i$ and $(x, y) = x + iy$.

We define $U_{(C)}$ to be the set of all ordered pairs (a, b) where $a, b \in U$, subject to the rules

$$(a, b) + (a', b') = (a + a', b + b'),$$

$$(x, y)(a, b) = (xa - yb, xb + ya),$$

and we identify U with the set of all ordered pairs $(a, 0)$ operated upon by the scalars $(x, 0)$. If we write $(a, b) = a + ib$, the second rule becomes

$$(x + iy)(a + ib) = xa - yb + i(xb + ya).$$

The reader will immediately verify that $U_{(C)}$ is in fact a vector space of dimension n over C. We shall refer to $U_{(C)}$ as the *complexification* of U.

EXERCISES

1. The vector space V over Q spanned by $1, i$ is a subspace of C, where C is regarded as a vector space over Q. In fact, V is even a subfield of C. (It is called the *field of Gaussian numbers*.)

2. Show that, in the notation used above, neither V_F nor U_F is a subspace of V_E ($E \neq F$).

3. Prove that an n-dimensional vector space over F_p is finite and consists of exactly p^n elements. Show further that the number of distinct ordered bases is $(p^n - 1)(p^n - p) \cdots (p^n - p^{n-1})$. How many distinct bases are there?

CHAPTER II

Affine and Projective Geometry

NOTE: *Except where we state the contrary, all vector spaces considered in the remainder of this book are assumed to be finite dimensional.*

2.1 Affine Geometries

In this chapter we shall introduce two different (but closely related) geometrical languages. The first of these, the language of affine geometry, is the one which appeals most closely to our intuitive ideas of geometry. In this language the subspaces of a vector space of dimensions 0, 1 and 2 are called "points", "lines" and "planes", respectively. But we cannot limit these words to describe only subspaces: otherwise V would have only one point, namely the zero subspace, and every line and plane in V would contain this point. Our intuition suggests that we introduce the concept of "translated" subspace.

DEFINITION. If a is any fixed vector in V and M is a subspace of V, then $a + M$ denotes the set of all vectors $a + m$ where m runs through M, and is called a *translated subspace* (or *coset*) in V.

For the sake of brevity we shall frequently use the word "coset" rather than the more descriptive term "translated subspace".

LEMMA 1. *The following statements are equivalent:*
(1) $a + M = b + M$;
(2) $b \in a + M$;
(3) $-a + b \in M$.

PROOF. If $a + M = b + M$, then $b = b + 0 \in a + M$. In other words, $-a + b = m$ for some m in M. Thus (1) implies (2) and (2) implies (3).

If now $-a + b = m \in M$, then $b + M = (a + m) + M = a + (m + M)$. But any vector in $m + M$ is of the form $m + m'$ where $m, m' \in M$ and so $m + M \subset M$. Similarly $-m + M \subset M$, which implies that $M \subset m + M$. Hence $M = m + M$ and we have shown that (3) implies (1).

As an immediate consequence we see that if the cosets $a + M$, $b + M$ have any vector c in common, then they coincide (with $c + M$).

LEMMA 2. *If $a + M = b + N$ then $M = N$.*

PROOF. If $a + M = b + N$, then $b = b + 0 \in a + M$ and so $a + M = b + M$ by Lemma 1. Thus $b + M = b + N$, from which the result follows.

We may thus say unambiguously that M is *the subspace belonging to the coset $a + M$.*

We base our construction of cosets on the following result.

PROPOSITION 1. *The intersection of any family of cosets in V is either a coset in V or is empty.*

PROOF. Let $(S_i)_{i \in I}$ be a given family of cosets and suppose that $c \in \cap S_i$. By Lemma 1, each S_i is of the form $c + M_i$ where M_i is the subspace belonging to S_i. Thus $\cap S_i = \cap (c + M_i) = c + \cap M_i$, which is a coset by Proposition 1 of Chapter I (p. 7).

DEFINITION. The *join* of a family $(S_i)_{i \in I}$ of cosets in V is the intersection of all the cosets in V which contain every S_i. We denote this join by $\mathsf{J}(S_i : i \in I)$, or simply by $\mathsf{J}(S_i)$.

For two cosets we more often use the notation $S_1 \mathsf{J} S_2$. Then clearly $(S_1 \mathsf{J} S_2) \mathsf{J} S_3 = S_1 \mathsf{J} (S_2 \mathsf{J} S_3) = \mathsf{J}(S_1, S_2, S_3)$. (See also exercise 1, below.)

It follows from Proposition 1 that $\mathsf{J}(S_i)$ is itself a coset in V and, of course, $\mathsf{J}(S_i)$ is the least coset in V containing every S_i.

Since a coset is a subspace if, and only if, it contains 0 (Lemma 1) we see that the definitions we have given of the join of subspaces (p. 7) and the join of cosets are consistent with each other.

The join of the family (S_i), as we have defined it, seems to depend on the containing vector space V, but if S is any coset in V which contains every S_i, $i \in I$, then $\mathsf{J}(S_i)$ is also the smallest coset *in S* containing every S_i. Thus $\mathsf{J}(S_i)$ could equally well be defined as the intersection of all the cosets *in S* containing every S_i, $i \in I$.

In view of Lemma 2 we may make the following definition.

DEFINITION. The *dimension* of the coset $a + M$ is the dimension of the subspace M and we write $\dim(a + M) = \dim M$.

DEFINITION. Let V be a vector space over F and let S be a coset in V. The set of all cosets in S is the *affine geometry* on S and will be denoted by $\mathscr{A}(S)$. The *dimension* of $\mathscr{A}(S)$, written $\dim \mathscr{A}(S)$, is $\dim S$. The elements of $\mathscr{A}(S)$ of dimensions 0, 1, 2 are called *points, lines, planes,* respectively, and the elements of $\mathscr{A}(S)$ of dimension $\dim S - 1$ are called *hyperplanes* in $\mathscr{A}(S)$ (or sometimes hyperplanes in S).

If the coset S is contained in the coset T then $\mathscr{A}(S)$ is said to be a *subgeometry* of $\mathscr{A}(T)$.

We note that according to our definition of affine geometry the points are sets, each consisting of just one vector. In other words, we distinguish between a vector v and the point $\{v\}$ that it determines. A little care with the notation is required if the reader thinks in terms of the intuitive concept of "position vector of a point". It is also well to remember that lines, as we have defined them, are not the same as "segments": they do not need to be "produced" and do not require the qualifying adjective "straight".

We shall frequently use the familiar notation P, Q, R, etc., for points, but reserve O for the point $\{0_v\}$. The line joining distinct points P, Q is written as PQ. We also use the adjectives "collinear", "concurrent", "coplanar", etc., in their usual sense.

The reader is encouraged to draw pictures or diagrams of the geometrical configurations he encounters. These are often an invaluable aid to thought but they must never, of course, be allowed to substitute as a proof!

EXERCISES

1. Given cosets $S_1, \ldots, S_n$, prove that every possible way of inserting parentheses in $S_1 \sqcup \cdots \sqcup S_n$ gives the same result, namely $\sqcup(S_1, \ldots, S_n)$.

2. If S is a coset, prove that $[S]$, the subspace spanned by S, is precisely $S \sqcup 0$.

3. Let V be a vector space over a field F and S a coset in V. If $a_1, \ldots, a_r$ lie in S and $x_1, \ldots, x_r$ are scalars such that $x_1 + \cdots + x_r = 1$, prove that $x_1 a_1 + \cdots + x_r a_r \in S$.

4. If a non-empty subset S of V has the property of exercise 3 for all r, show that S is a coset in V.

5. If F has at least three elements and S is a non-empty subset of V satisfying the condition of exercise 3 for $r = 2$, show again that S is a coset. By considering $\mathscr{A}(\mathbf{F}_2{}^2)$ show that this condition on the field F is necessary.

6. Prove that the set of all solutions (x, y, z) in $\mathbf{R}^3$ of the linear equations $X - Y + Z = 1$, $X + Y + 2Z = 2$, is a coset in $\mathbf{R}^3$ of dimension 1.

2.2 Affine Propositions of Incidence

In the previous chapter we found a simple relationship between the dimensions of the intersection and join of two subspaces (Theorem 2 (2)). In the case of cosets, however, the situation is complicated by the fact that their intersection may be empty. We have the following criterion.

PROPOSITION 2. *Let S and T be cosets in V and M, N the subspaces belonging to S, T respectively. Then $S \cap T$ is empty if, and only if,*

$$\dim(S \sqcup T) = \dim(M + N) + 1.$$

This result is an immediate consequence of the following two lemmas.

LEMMA 3. $(a+M) \cup (b+N) = a + [-a+b] + M + N.$

PROOF. $(a+M) \cup (b+N)$ is a coset containing a and b and so is of the form $a+P = b+P$ for some subspace P (Lemma 1). Then P certainly contains $[-a+b]$ (Lemma 1), and also M and N. But the join of $a+M$, $b+N$ is the least coset containing them both. Hence

$$P = [-a+b] + M + N.$$

LEMMA 4. $(a+M) \cap (b+N)$ *is not empty if, and only if,* $-a+b$ *belongs to* $M+N$.

PROOF.   We have $a+m = b+n$ if, and only if, $-a+b = m-n$.

In order to express the fundamental propositions of incidence in affine geometry we need a more delicate notion than that of empty intersection. For example, in the intuitive three dimensional geometry with which we are familiar, the join of two lines without common points is a plane if the lines are "parallel" and is the whole space if the lines are "skew".

DEFINITION.   The cosets $a+M$, $b+N$ are said to be *parallel* if

$$M \subset N \text{ or } N \subset M.$$

Note that our definition implies that a point is parallel to each coset.

It is clear from this definition that if the coset S is contained in the coset T then S and T are parallel (argue as in the proof of Lemma 2). On the other hand, if S and T are parallel and neither one contains the other, then they have no points in common. For, if $c \in S \cap T$ then $S = c+M$, $T = c+N$ and so $M \subset N$ or $N \subset M$ implies $S \subset T$ or $T \subset S$, respectively.

A convenient method of indicating in a diagram that two lines are parallel is to draw them with arrow-heads (though the direction of the arrows is immaterial).

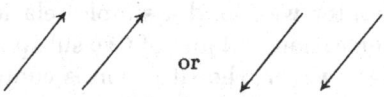

Figure 1.   Parallel lines

The fundamental result which incorporates all the propositions of incidence in affine geometry can now be stated.

THEOREM 1. *Let S and T be cosets in the vector space V.*
(1) *If $S \subset T$, then* $\dim S \leq \dim T$ *and* $\dim S = \dim T$ *implies* $S = T$.
(2) *If $S \cap T$ is not empty, then*

$$\dim(S \cup T) + \dim(S \cap T) = \dim S + \dim T.$$

(3) *Assuming $S \cap T$ is not empty, S and T are parallel if, and only if, one contains the other.*
Assuming $S \cap T$ is empty, S and T are parallel if, and only if,

$$\dim(S \cup T) = \max(\dim S, \dim T) + 1.$$

PROOF. Let M and N be the subspaces belonging to S and T respectively.

In parts (1) and (2) we choose c in $S \cap T$ and then $S = c + M$, $T = c + N$, $S \cup T = c + (M + N)$, $S \cap T = c + (M \cap N)$. Both results now follow from Theorem 2 of Chapter I (p. 11).

We have already proved the first part of (3). On the assumption that $S \cap T$ is empty, Proposition 2 above and Theorem 2 (1) of Chapter I show that $\dim(S \cup T) = \max(\dim M, \dim N) + 1$ if, and only if, $M + N = N$ or $M + N = M$, i.e., if, and only if, $M \subset N$ or $N \subset M$.

We now state the propositions of incidence in dimensions two and three. There are, of course, propositions of incidence for affine geometries of all dimensions and their proofs depend only on Theorem 1. We shall merely establish here two of the following propositions and leave the reader to prove the others.

Propositions of Incidence in $\mathscr{A}(V)$, where $\dim V = 2$.
2.1   *The join of two distinct points is a line.*
2.2   *The intersection of two non-parallel lines is a point.*

Propositions of Incidence in $\mathscr{A}(V)$, where $\dim V = 3$.
3.1   *The join of two distinct points is a line.*
3.2   *The intersection of two non-parallel planes is a line.*
3.3   *The join of two lines with a point of intersection is a plane.*
3.4   *The intersection of two coplanar non-parallel lines is a point.*
3.5   *The join of two distinct parallel lines is a plane.*
3.6   *The join of a point and a line not containing it is a plane.*
3.7   *The intersection of a plane and a line not parallel to it is a point.*

PROOF OF 2.2: Let S, T be non-parallel lines in the affine plane $\mathscr{A}(V)$. Since $S \neq T$, $\dim(S \cup T) > \dim S = 1$ by Theorem 1 (1) (applied to S and $S \cup T$), and hence $\dim(S \cup T) = 2$ (because $\dim V = 2$). Now $S \cap T$ cannot be empty, otherwise S and T would be parallel by Theorem 1 (3), and so by Theorem 1 (2) $\dim(S \cap T) = 1 + 1 - 2 = 0$.

PROOF OF 3.5. If S, T are distinct parallel lines, then $S \cap T$ is empty and $\dim(S \cup T) = \max(1, 1) + 1 = 2$.

We shall say that two lines S and T are *skew* if they have empty intersection and are not parallel. Thus, by 2.2, skew lines cannot be coplanar.

EXERCISES

1. Show that the affine geometry $\mathscr{A}(\mathbf{F}_2{}^2)$ has just four points and six lines (parallel in pairs).

2. If S, T are skew lines in an affine geometry of dimension three, show that there is a unique plane P_S containing S and parallel to T, and a unique plane P_T containing T and parallel to S. Show also that P_S and P_T are parallel.

3. If S, T are two cosets and if K is the set of all $xs + yt$ where x, y are scalars such that $x + y = 1$, $s \in S$ and $t \in T$, show that K contains every line joining a point of S to a point of T. In the case of exercise 2 above, show that K is not a coset. (This explodes what looks like a plausible definition of the join of cosets "from below".)

4. If $a_1, \ldots, a_r$ are vectors in V, prove that
$$\mathsf{J}(a_1, \ldots, a_r) = a_1 + [a_i - a_1; i = 2, \ldots, r].$$
Deduce that $\mathsf{J}(a_1, \ldots, a_r)$ has dimension $r - 1$ if, and only if, $a_i - a_1$, $i = 2, \ldots, r$, are linearly independent.

2.3 Affine Isomorphisms

If S, T are cosets of the same dimension in a vector space V, then there is an intuitively obvious sense in which the affine geometries $\mathscr{A}(S)$, $\mathscr{A}(T)$ are equivalent: any construction or result that is possible in the one is automatically also valid in the other.

DEFINITION. If A, A' are affine geometries, and α is a one–one mapping of A onto A' such that
$$S \subset T \quad \text{if, and only if,} \quad S\alpha \subset T\alpha$$
for all S, T in A, then α is called an *isomorphism* of A onto A'. If such an isomorphism exists, we say A, A' are *isomorphic* (or that A is isomorphic to A').

We may express the condition on α in the definition of isomorphism by saying that α and α^{-1} *preserve inclusion*. But if our intuitive idea of isomorphism is correct then we shall expect an isomorphism to preserve not only inclusions but also intersections, joins, dimensions and parallelism. This we state more formally as follows.

PROPOSITION 3.   *Let* A, A' *be affine geometries and* α *an isomorphism of* A *onto* A'.

If $(S_i)_{i \in I}$ *is a family of elements of* A, *then*
(1) $(\cap S_i)\alpha = \cap(S_i\alpha)$ *if* $\cap S_i$ *is not empty, and*
(2) $\mathsf{J}(S_i)\,\alpha = \mathsf{J}(S_i\alpha)$.
Further,
(3) dim $S\alpha = $ dim S, *for each* S *in* A, *and*
(4) *the elements* S, T *of* A *are parallel if, and only if,* $S\alpha$, $T\alpha$ *are parallel.*

PROOF.   The coset $\cap S_i = C$ is the largest coset contained in every S_i, $i \in I$: this means that C is the coset uniquely determined by the two conditions (i) $C \subset S_i$ for each i in I, and (ii) if $T \subset S_i$ for each i in I, then $T \subset C$. It follows that $C\alpha$ is the largest coset contained in every $S_i\alpha$, $i \in I$, i.e., $C\alpha = \cap S_i\alpha$. This has established (1).   Part (2) follows by a similar argument since $\mathsf{J}(S_i)$ is the smallest coset containing every S_i, $i \in I$.

For part (3) we consider chains of the form

$$C = C_0 \supset C_1 \supset \cdots \supset C_k$$

where each C_i is a coset and $C_{i+1} \neq C_i$ for $i = 0, \ldots, k-1$.   The length of this chain is k and one sees immediately from Theorem 1 (1) that the length of the longest such chain is precisely dim C.

Part (4) is now a direct consequence of Theorem 1 (3) and parts (1), (2) and (3).

We remark that the above proof serves also to show that the intersection and join of cosets, as well as coset dimension and parallelism can all be expressed solely by the use of the inclusion relation $\subset$ on cosets.

There is an analogue for affine geometries of Theorem 3 of Chapter I.

THEOREM 2.   *Any two affine geometries of the same dimension over the same field are isomorphic.*

PROOF.   Let A $= \mathscr{A}(a + M)$, A' $= \mathscr{A}(a' + M')$ where $a + M$, $a' + M'$ are cosets of the same dimension in vector spaces V, V'. Theorem 3 of Chapter I shows that there is an isomorphism f of M onto M'. The mapping

$$\alpha: T \to (T-a)f + a',$$

where $T \in$ A, carries A onto A' and has inverse mapping

$$\alpha^{-1}: T' \to (T'-a')f^{-1} + a,$$

where $T' \in A'$. Since α and α^{-1} clearly preserve inclusion, α is an isomorphism of A onto A'. (We shall return to a study of these isomorphisms in the next chapter.)

In view of this theorem we shall often speak of *"affine geometry of dimension n (over F)"*.

Also, the fact that F^{n+1} contains cosets of dimension n which do not contain 0 shows that we can find an isomorphic copy of any affine geometry of dimension n over F in another affine geometry $\mathscr{A}(V)$ and so that the copy does not contain 0_V.

There is an important point concerning our definition of isomorphism for affine geometries which has so far remained in the background. To say that A is an affine geometry means that $A = \mathscr{A}(S)$ for some coset S in some vector space V over a field F. If A' is another affine geometry, then $A' = \mathscr{A}(S')$ where S' is a coset in a vector space V' over a field F'. Even if A and A' are isomorphic there is no a priori reason for F' to coincide with F. If, for example, $\dim_F V = \dim_{F'} V' = 1$ and α is *any* one–one mapping of F onto F', then α gives rise to an isomorphism of $\mathscr{A}(V)$ onto $\mathscr{A}(V')$. We shall see in the next chapter (§ 3.5) that, in higher dimensions, if $\mathscr{A}(V)$ and $\mathscr{A}(V')$ are isomorphic, then the fields F and F' are not only related by a one–one correspondence but have exactly the same algebraic structure.

2.4   Homogeneous Vectors

We shall prove some fundamental theorems of affine geometry by means of what are called "homogeneous vectors". The method is really one of the tools of projective geometry (to be introduced shortly), but we hope that our present use of it will serve to pave the way for the less familiar notions of projective geometry.

Let S be a plane in an affine geometry $\mathscr{A}(V)$ of dimension 3 over F, and suppose S does not contain the zero vector 0_V. If P is a point on S, then any *non-zero* vector p in OP will be called a *homogeneous vector* for P. In other words, $OP = [p]$. The mapping $P \to OP$ is a one–one mapping of S onto the set of all lines through O not parallel to S. Further, three points P, Q, R of S are collinear if, and only if, OP, OQ, OR are coplanar. This follows at once from the propositions of incidence 3.2 and 3.6.

LEMMA 5.   *Given three distinct collinear points P, Q, R in S, a homogeneous vector p for P, and any two non-zero scalars x, y, then there exist homogeneous vectors q, r for Q, R, respectively, such that $p = xq + yr$.*

PROOF. If q_0, r_0 are any homogeneous vectors for Q, R then p, q_0, r_0 are linearly dependent since OP, OQ, OR are coplanar. But q_0, r_0 are linearly independent since $OQ \neq OR$. Hence, by Proposition 2 (2) of Chapter I (p. 9), $p = x_0 q_0 + y_0 r_0$ for some non-zero scalars x_0, y_0. We now put $q = (x_0/x)\, q_0$ and $r = (y_0/y)\, r_0$.

Lemma 5 allows us to simplify the constants appearing in arguments using homogeneous vectors.

DEFINITION. If A, B, C are three non-collinear points then the configuration consisting of these three points and their three joins is called the *triangle* ABC.

We shall say that the two triangles ABC, $A'B'C'$ are in *perspective from a point* P if the seven points P, A, B, C, A', B', C' are distinct and if the three lines AA', BB', CC' are distinct and meet at P. We then refer to P as the *center of perspective* of the triangles.

THEOREM 3. (DESARGUES' THEOREM FOR THE AFFINE PLANE.) *If* ABC, $A'B'C'$ *are two coplanar triangles in perspective from a point* P, *and if the pairs of corresponding sides intersect in points* L, M, N, *then* L, M, N *are distinct and collinear and the line* LMN *is distinct from the six sides of the triangles.* (*Cf. Fig. 2*)

PROOF. We may assume that the plane S of the triangles ABC, $A'B'C'$ is contained in an affine geometry $\mathcal{A}(V)$ but does not contain 0_V.

We choose any homogeneous vector p for P and use Lemma 5 to find homogeneous vectors a, a'; b, b'; c, c' for A, A'; B, B'; C, C', respectively such that

$$a' = p + a, \quad b' = p + b, \quad c' = p + c.$$

Then $b' - c' = b - c = l$, say, where $l \neq 0$ since $OB \neq OC$. Now l lies in the planes OBC, $OB'C'$ and so lies in the line OL where $L = BC \cap B'C'$, i.e., l is a homogeneous vector for L. Similarly $c' - a' = c - a = m$ is a homogeneous vector for M, and $a' - b' = a - b = n$ is a homogeneous vector for N. But $l + m + n = 0$ and so OL, OM, ON are coplanar. This implies that L, M, N are collinear. It follows at once from the linear independence of a, b, c that L, M, N are distinct and none of the points A, B, C lies on the line LMN.

The simplicity of this proof is due to the use of homogeneous vectors. But this is not the main reason for employing them. Their real significance is that their use in the proof points the way to further results not included in the original version of the theorem. For example, if AA', BB', CC' are parallel instead of being concurrent, then we may still take a non-zero vector p in the line through O parallel to AA', BB',

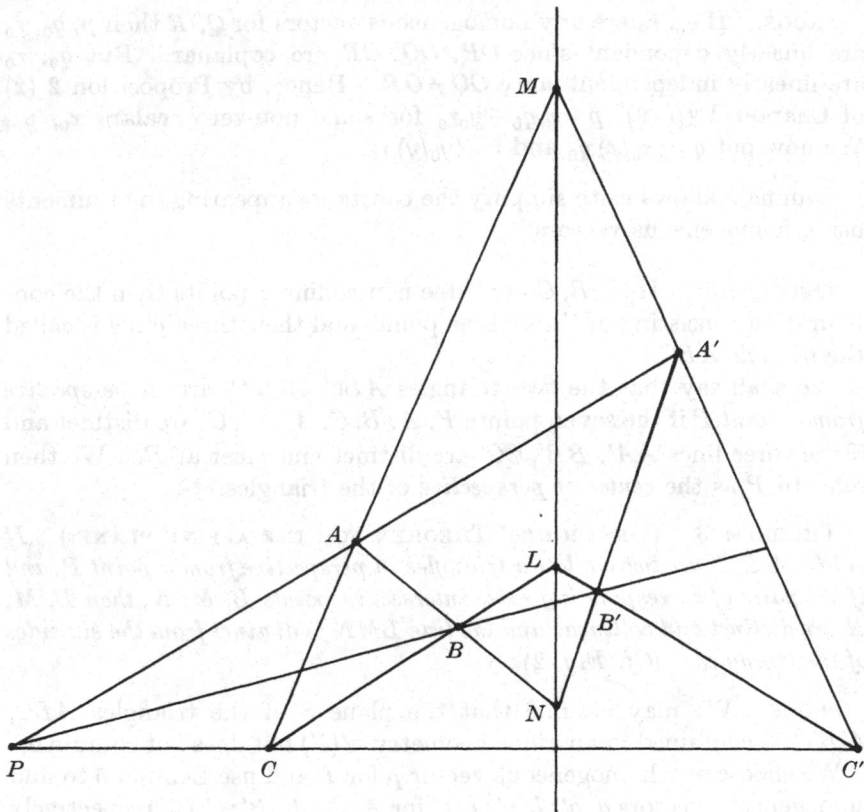

Figure 2

CC' and argue precisely as before. In proving Lemma 5 we only used the fact that OP, OQ, OR are coplanar. In this case we use the fact that p, OA, OA' are coplanar. The conclusion of the theorem is unaltered. In these circumstances we shall still say that the triangles are in perspective. This configuration is illustrated in Figure 3. We have labelled the arrow-heads (of the parallel lines) with the letter P to remind us that in the general configuration (cf. Fig. 2) these three lines meet in P. Here the point P has been "pushed to infinity".

Again, if BC is parallel to $B'C'$ and CA is parallel to $C'A'$, there are no points L, M, but the non-zero vectors l, m, n exist just as before. In this case $[l]$, $[m]$ are parallel to S and so $[n]$, which lies in the plane of $[l]$ and $[m]$ (since $l + m + n = 0$), is also parallel to S. This shows that AB is parallel to $A'B'$.

The two results we have just deduced are important affine theorems

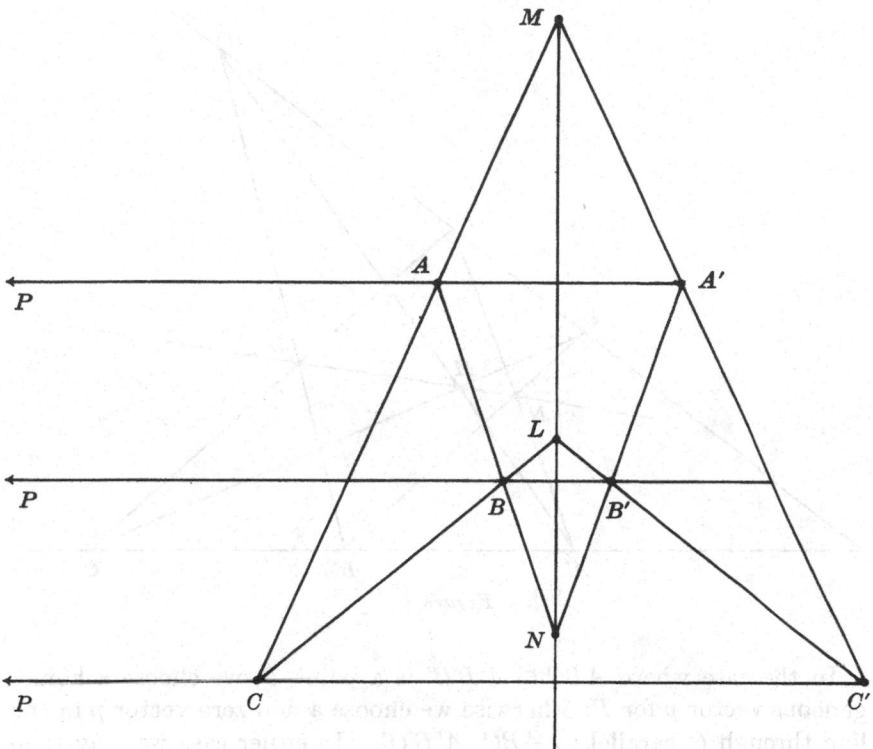

Figure 3

which cannot be inferred directly from the affine form of Desargues'
Theorem.   Their place will be clarified as soon as we establish the connexion between affine and projective geometry.

THEOREM 4.   (PAPPUS' (AFFINE) THEOREM.)   *Two triads of points*
A, B, C; A′, B′, C′ are taken on two distinct coplanar lines (possibly
parallel).   If the intersections $BC′ \cap B′C, CA′ \cap C′A, AB′ \cap A′B$ are
points L, M, N, respectively, then L, M, N are collinear.   (Cf. Fig. 4)

PROOF.   We remark first that the points of a triad are, by definition,
distinct.   If the lines $ABC, A′B′C′$ intersect in a point P, we shall assume in our proof that the seven points $P, A, B, C, A′, B′, C′$ are distinct.
But the reader will find it trivial to check that the theorem is also true
without this restriction.

The plane S containing ABC and $A′B′C′$ may be assumed to lie in an
affine geometry $\mathscr{A}(V)$ but not to contain 0_V.

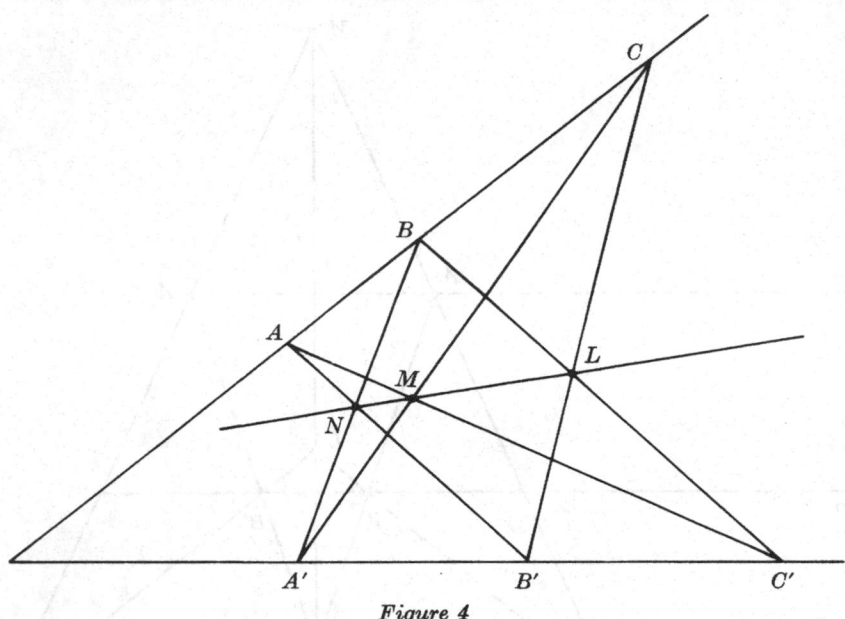

Figure 4

In the case where $ABC \cap A'B'C'$ is a point P, we choose a homogeneous vector p for P; otherwise we choose a non-zero vector p in the line through O parallel to ABC, $A'B'C'$. In either case we may then find homogeneous vectors a, b, a', b' for A, B, A', B' respectively so that $b = p + a$ and $b' = p + a'$ (Lemma 5). We may also find homogeneous vectors c, c' for C, C' respectively of the form $c = p + xa$, $c' = p + ya'$, where x, y are scalars.

Now $p + a + a' = n$, say, is a linear combination of a and b', and also of a' and b (and therefore cannot be 0, since $OA \neq OB'$). Hence n lies in the planes OAB', $OA'B$ and is a homogeneous vector for N. Similarly $p + xa + ya' = m$ is a homogeneous vector for M. Finally,

$$xyn - m = x(y-1)b + (x-1)c',$$

and

$$yxn - m = y(x-1)b' + (y-1)c.$$

Since $xy = yx$ for all scalars x, y, it follows that $l = xyn - m$ is a homogeneous vector for L. The linear dependence $l + m - xyn = 0$ shows that OL, OM, ON are coplanar and thus L, M, N are collinear.

The method of proof gives the further result that if BC' is parallel to $B'C$ and if CA' is parallel to $C'A$, then AB' is parallel to $A'B$. (Draw your own picture!)

THEOREM 5. (THE HARMONIC CONSTRUCTION IN THE AFFINE PLANE.) *Given two distinct points A, B in an affine plane S, and a point G on AB. We choose any point C in S but not on AB and also any point D on GC distinct from G and C. If $E = AD \cap BC$, $F = BD \cap CA$ and $H = EF \cap AB$ are points, then H is independent of the choice of C and D.*

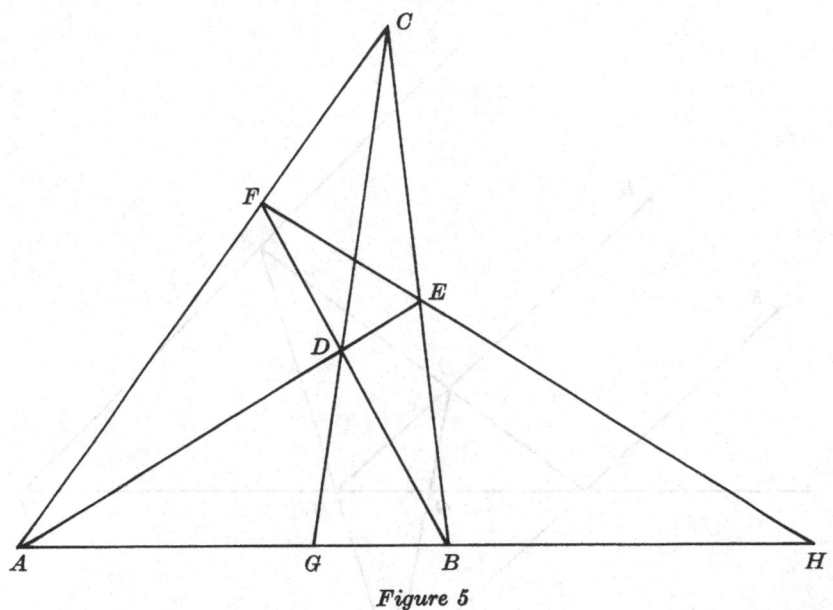

Figure 5

PROOF. As before we assume that S is contained in an affine geometry $\mathscr{A}(V)$ and does not contain 0_V. We then take homogeneous vectors, a, b, c, g for A, B, C, G respectively. Since OA, OB, OG are coplanar we have $g = xa + yb$, for some scalars x, y. We may then take a homogeneous vector d for D of the form $g + zc = xa + yb + zc$, for some scalar z. If we now put $e = yb + zc = d - xa$ and $f = xa + zc = d - yb$, we see immediately that e, f are non-zero vectors and are homogeneous vectors for E, F respectively. But then $h = xa - yb$ is a homogeneous vector for H. Comparing g and h gives the result.

DEFINITION. The point H of Theorem 5 is called the *harmonic conjugate* of G with respect to A and B. We also express the relationship between the four points by saying that $(A, B; G, H)$ is a *harmonic range*.

Again we have a situation where the proof yields further results. It is assumed in the statement of Theorem 5 that the points E, F, H all exist: i.e., that none of the line pairs AD, BC; AC, BD; EF, AB are parallel. But for certain choices of C and D some of these pairs may indeed be parallel. Nevertheless, the vectors e, f, h *always* exist—thus $[e]$ is always the intersection of the planes OAD, OBC, etc. Our proof therefore yields a construction for H even in certain situations not covered by the statement of the theorem. Let us interpret these cases in geometrical language.

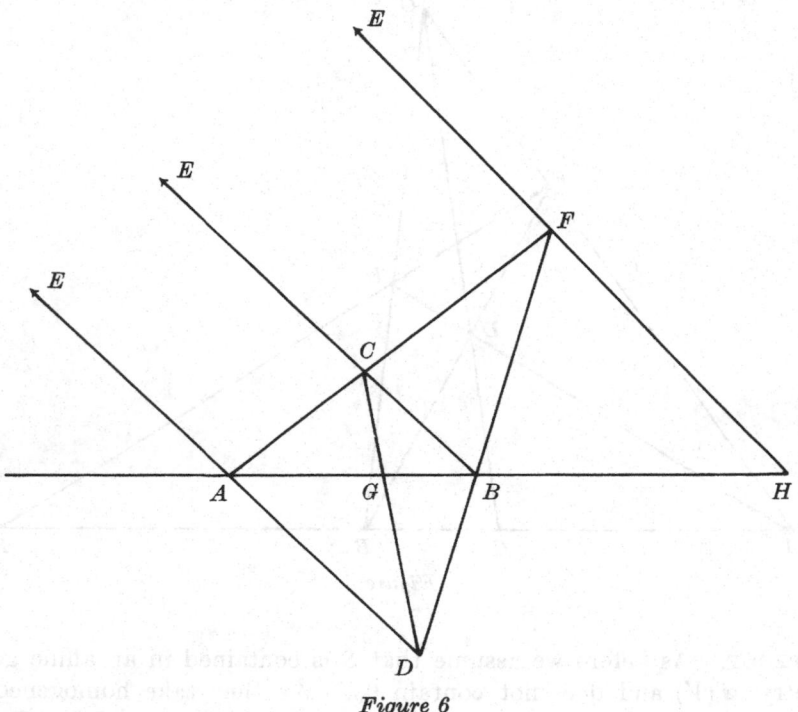

Figure 6

Suppose that AD is parallel to BC but that $BD \cap CA$ is a point F. Then our proof supplies H as the intersection (if it exists) of AB with the line through F parallel to AD (cf. Fig. 6); similarly, if E exists but not F. The vector h fails to give a point H on AB if, and only if, $[h]$ is the line through O parallel to AB.

To examine this construction in more detail it is convenient to revert to the situation where a, b are the actual vectors (and not just homogeneous vectors) for the points A, B. In other words we assume that $A = \{a\}$ and $B = \{b\}$. The points of AB are then precisely all the points of the form $\{xa + yb\}$ where x, y belong to the underlying field F and $x + y = 1$ (cf. Lemma 3 (p. 18) in the case $M = N = 0$). We refer to $\{xa + yb\}$, where $x + y = 1$, as *the point dividing A, B in the ratio $y:x$.*

(Notice the inverted order of x and y in this ratio.)   In particular, if the underlying field contains $\frac{1}{2}$, we refer to $\{\frac{1}{2}(a+b)\}$ as the *mid-point* of A, B.

Now suppose that $G = \{xa + yb\}$ where $x + y = 1$. Then the harmonic conjugate $H = \{h\}$ is given by $h = (xa - yb)/(x - y)$ provided only that $x \neq y$. Thus every point G of AB, excluding the mid-point (if it exists), possesses a unique harmonic conjugate H, and we have another classical definition of the harmonic range in which "G and H divide A, B internally and externally in the same ratio". Of course, the ideas of "internal" and "external" do not make sense unless the field F has some notion of order and so it is only safe for us to use this notion of subdivision if F is $\mathbf{Q}$ or $\mathbf{R}$.

It is tacitly assumed in Theorems 3, 4 and 5 that the affine geometries in question contain enough points for the configurations they describe to exist. An affine plane over $\mathbf{F_2}$ contains only four points and so we must obviously assume that the ground field contains at least three elements. This will give enough points for all three theorems provided we allow parallel lines. For the general cases we must assume that the ground field contains at least four elements.

EXERCISES

1. If H is the harmonic conjugate of G with respect to A, B we write $H = (A, B):G$. Show then that $G = (A, B):H$ and also, if $G \neq H$, that $B = (G, H):A$.

2. Show that $(A, B):A = A$ and $(A, B):B = B$.

3. Prove that, in the notation of Theorem 5, $G = H$, if, and only if, $2xy = 0$.

4. Prove that in an affine plane over $\mathbf{F_3}$, each line contains three distinct points and that each of these is the mid-point of the other two.

5. Let ABC be a triangle in $\mathscr{A}(\mathbf{R}^2)$ and denote by A', B', C', respectively, the mid-points of B, C; C, A; A, B. Show that the "medians" AA', BB', CC' are concurrent. If the same configuration is taken in $\mathscr{A}(\mathbf{F_3}^2)$, show that the medians are parallel.

2.5 Projective Geometries

The incidence properties of affine geometry forcefully suggest that we need some notion of "points at infinity". The gap is filled by a new concept, that of projective geometry.

DEFINITION.   Let V be a vector space over F. The set of all subspaces of V is the *projective geometry* on V and will be denoted by $\mathscr{P}(V)$.

The relationship between affine and projective geometries is stated in Theorem 7, at the beginning of the next section. The reader is urged to look now at the statement of that theorem and in particular at part (6) since he will then appreciate the motivation for the following definition.

DEFINITION. If M is a subspace of a vector space V, the *projective dimension* of M is defined to be dim $M - 1$ and will be denoted by pdim M. The *dimension* of the geometry $\mathscr{P}(V)$, written pdim $\mathscr{P}(V)$, will be pdim V. The elements of $\mathscr{P}(V)$ of projective dimensions 0, 1, 2 are called projective points, projective lines, projective planes, respectively. An element of $\mathscr{P}(V)$ of projective dimension pdim $V - 1$ is called a *hyperplane* in $\mathscr{P}(V)$. (It is necessarily a hyperplane in V as previously defined, and also a subspace.)

If M is a subspace of V, then $\mathscr{P}(M)$ is called a *subgeometry* of $\mathscr{P}(V)$. In particular, if M is a projective point, then $\mathscr{P}(M)$ consists only of the two elements 0, M. Moreover, the point M is determined by any non-zero vector m in M, i.e., $[m] = M$: such a vector is called a *homogeneous vector* for M.

In many geometrical considerations there is a natural emphasis on the concept of point. Curves and surfaces, for example, are usually thought of as consisting of the points lying on them. In these circumstances the higher dimensional elements of our projective geometry, the lines, planes, etc., are also considered as sets of points. To make matters precise we introduce a definition.

DEFINITION. The *projective space* determined by the projective geometry $\mathscr{P}(V)$ is the set of all the projective points in $\mathscr{P}(V)$.

Each non-zero element T of $\mathscr{P}(V)$ gives rise to a subset of the projective space, namely the set of all projective points in T. But the zero subspace (of projective dimension -1) contains no projective points and hence yields the empty subset of projective space. Thus, for example, if two projective lines M, N are such that $M \cap N = 0$, then M, N have no projective points in common. More generally, two elements M, N of $\mathscr{P}(V)$ have no projective points in common if, and only if, $M \cap N = 0$, or equivalently, if, and only if, pdim $(M \cap N) = -1$. Such elements of $\mathscr{P}(V)$ are called *skew*.

Throughout the rest of this book, when it is clear from the context that we are discussing projective geometry, we shall omit the adjective "projective" before the words point, line and plane.

As in affine geometry, we shall frequently write P, Q, R, etc. for points; PQ for the line joining the distinct points P, Q; and use the familiar terms "collinear", "concurrent", "coplanar", etc., in their natural meaning.

For comparison with the affine case we state below the propositions of incidence in projective geometries of dimensions two and three.

Propositions of Incidence in $\mathscr{P}(V)$, where pdim $V = 2$.

2.1 *The join of two distinct points is a line.*

2.2 *The intersection of two distinct lines is a point.*

Propositions of Incidence in $\mathscr{P}(V)$, where pdim $V = 3$.

3.1 *The join of two distinct points is a line.*

3.2 *The intersection of two distinct planes is a line.*

3.3 *The join of two distinct intersecting lines is a plane.*

3.4 *The intersection of two distinct coplanar lines is a point.*

3.5 *The join of a point and a line not containing it is a plane.*

3.6 *The intersection of a plane and a line not contained in it is a point.*

The reader will see at once that the statements of these propositions are much simpler than those for the corresponding affine propositions although, as we shall see (in § 2.6) the latter may be deduced from the former. The proofs are also much simpler as they all follow from Theorem 2 of Chapter I. For example, 2.2: We are given distinct lines M and N so that pdim $M =$ pdim $N = 1$ and pdim$(M + N) > 1$. Hence pdim$(M + N) = 2$ and consequently

$$\text{pdim}(M \cap N) = \text{pdim } M + \text{pdim } N - \text{pdim}(M + N) = 0.$$

We have seen, in connexion with affine geometry, how to express joins, intersections and dimensions of cosets by using only the inclusion relation $\subset$ on the cosets (see the remark immediately following Proposition 3). The proofs of these results apply equally well when restricted to subspaces. We conclude that sums, intersections and projective dimensions of subspaces are all definable solely in terms of $\subset$.

The notion of an isomorphism of projective geometries is defined exactly as for affine geometries:

DEFINITION. A one–one mapping π of a projective geometry P onto a projective geometry P$'$ is called an *isomorphism* if it satisfies the condition that

$$M \subset N \quad \text{if, and only if,} \quad M\pi \subset N\pi$$

for all M, N in P. If such an isomorphism exists, we say that P and P$'$ are *isomorphic* (or that P is isomorphic to P$'$).

It is clear that an isomorphism of projective geometries preserves intersections, sums and projective dimensions. There is again an analogue of Theorem 3 of Chapter I.

THEOREM 6. *Any two projective geometries of the same dimension over the same field are isomorphic.*

For if f is an isomorphism of V onto V', then the mapping

$$\pi \colon M \to Mf$$

is clearly an isomorphism of $\mathscr{P}(V)$ onto $\mathscr{P}(V')$.

It follows that we may talk of *"projective geometry of dimension n (over F)"*.

EXERCISES

1. If P is a point in a three dimensional projective geometry and L, M are skew lines not containing P, prove that there exists a unique line through P intersecting L and M.

2. If $\mathscr{P}(V)$ is a four dimensional projective geometry and L, M, N are mutually skew lines not lying in a hyperplane in $\mathscr{P}(V)$, prove that L intersects $M + N$ in a point. Hence show that there is a unique line in $\mathscr{P}(V)$ intersecting L, M and N.

3. Let P be a point and L a line in a given projective geometry $\mathscr{P}(V)$. If P does not lie on L show that $Q \to PQ$, where Q is a point of L, is a one–one mapping of the points of L onto the lines in $P + L$ that go through P.

4. Prove that every line in $\mathscr{P}(\mathbf{F}_p{}^3)$ has $p + 1$ points on it. Deduce from this that the total number of points in $\mathscr{P}(\mathbf{F}_p{}^3)$ is $p^2 + p + 1$. How many lines are there? (Use exercise 3.)

5. Let V be an $n + 1$-dimensional vector space over $\mathbf{F}_p$. If u, v are non-zero vectors, write $u \sim v$ whenever $[u] = [v]$. Prove that $\sim$ is an equivalence relation on the set of non-zero vectors and deduce that the number of points in $\mathscr{P}(V)$ is $1 + p + \cdots + p^n$.

6. Prove the following theorem in projective geometry of dimension 3, using only the propositions of incidence: If ABC, $A'B'C'$ are two triangles, *in distinct planes*, which are in perspective from a point P, then the pairs of corresponding sides BC, $B'C'$; CA, $C'A'$; AB, $A'B'$ intersect in collinear points.

2.6   The Embedding of Affine Geometry in Projective Geometry

The fundamental connexion between affine and projective geometry is given in

THEOREM 7. (THE EMBEDDING THEOREM.) *If H is any hyperplane in $\mathscr{P}(V)$ and c is any vector of V not in H, then the mapping*

$$\varphi \colon S \to [S]$$

of $\mathscr{A}(c+H)$ into $\mathscr{P}(V)$ has the following properties:

(1) *φ is a one–one mapping;*

(2) *the image of $\mathscr{A}(c+H)$ under φ is the subset A of $\mathscr{P}(V)$ consisting of all subspaces of V which are not contained in H;*

(3) *$S \subset T$ if, and only if, $[S] \subset [T]$ for all S, T in $\mathscr{A}(c+H)$;*

if $(S_i)_{i \in I}$ is a family of cosets in $c+H$, then

(4) *$[\cap S_i] = \cap[S_i]$ if $\cap S_i$ is not empty, and*

(5) *$[\mathsf{J}(S_i)] = +[S_i]$;*

(6) *dim $S = $ pdim $[S]$ for all cosets S in $c+H$;*

(7) *S and T are parallel in $\mathscr{A}(c+H)$ if, and only if, one of the subspaces $[S] \cap H$, $[T] \cap H$ contains the other.*

PROOF. (1) If $S = a + M \in \mathscr{A}(c+H)$, then $[S] = [a] + M$ and $c+H = a+H$ (by Lemma 1). If $xa + m = a + h$, then $(x-1)a = h - m \in H$ where $a \notin H$ (since $c \notin H$), and so $x = 1$ and $h = m$. Hence $[S] \cap (c+H) = S$. This shows that the mapping φ is one–one and also, incidentally, that the inverse of φ is the mapping $S\varphi \rightarrow (S\varphi) \cap (c+H)$.

(2) If P is a subspace of V not contained in H, then $P + H = V$, since H is a hyperplane in V, and so

$$\dim(P \cap H) = \dim P + \dim H - \dim V = \dim P - 1.$$

In other words, $P \cap H$ is a hyperplane in P.

Let t be a vector in P, not in H. Then it follows immediately by counting dimensions that $P = [t] \oplus (P \cap H)$. Since $H + [t] = V$, it follows in particular that c is of the form $yt + h$ and so $yt \in c + H$. We put $a = yt$ and $M = P \cap H$. Then $P = [a] + M$, where $a + M \in \mathscr{A}(c+H)$. This shows that φ is onto A.

(3) If $[S] \subset [T]$ then $[S] \cap (c+H) \subset [T] \cap (c+H)$, which is precisely $S \subset T$ by the last remark in the proof of (1).

(4) If $e \in \cap S_i$ then $S_i = e + M_i$, for each i in I. Hence

$$\cap S_i = e + \cap M_i \quad \text{and thus} \quad [\cap S_i] = [e] + \cap M_i.$$

Further,

$$\cap[S_i] = \cap([e] + M_i) = [e] + \cap M_i,$$

for if $xe + m_i = ye + m_j$, then $(x-y)e = m_j - m_i \in H$, so that $x = y$ and $m_i = m_j$.

(5) $[\mathsf{J}(S_i)] = \mathsf{J}(S_i) \mathsf{J} 0 = \mathsf{J}(S_i \mathsf{J} 0) = \mathsf{J}([S_i])$, since each of these subspaces is the least coset containing 0 and S_i for each i in I; and

$$\mathsf{J}([S_i]) = +[S_i],$$

as J and $+$ are equivalent for subspaces.

(6) If $S = a + M$, then dim $[S] = $ dim $M + 1$, since $[a] \cap M = 0$. Hence pdim $[S] = $ dim S.

(7) If $S = a + M$, then $[S] \cap H = M$. For if $xa + m = h$ we must have $x = 0$, $m = h$. The result now follows at once from the definition of parallel cosets.

The above proof can easily be visualized in the case when dim $V = 3$. Figure 7 illustrates the connexion between a point $\{a\}$ of $\mathscr{A}(c + H)$ and its image $[a]$ in $\mathscr{P}(V)$; while Figure 8 shows the images of two parallel lines.

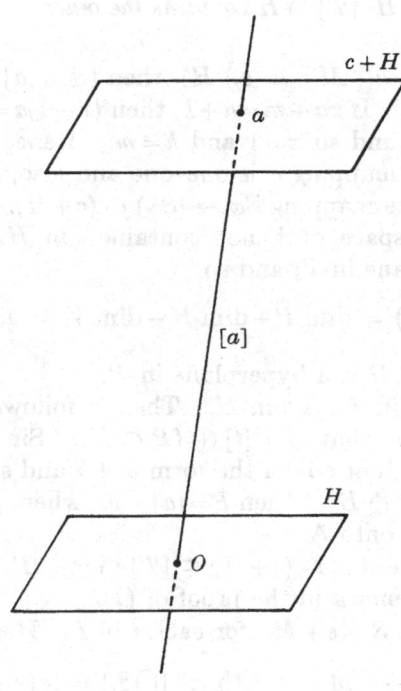

Figure 7

Theorem 7 illustrates a situation frequently encountered in geometry. The set A is linked by a one–one mapping φ to the geometry $\mathscr{A}(c + H)$. All the geometrical notions present in $\mathscr{A}(c + H)$, such as inclusion, inter-section, join, dimension and parallelism, can be carried over to A by means of φ. In this way A itself may be regarded as an affine geometry and there are, in fact, many situations where it is more fruitful to focus attention on A rather than on $\mathscr{A}(c + H)$. We are thus led to the follow-ing definition.

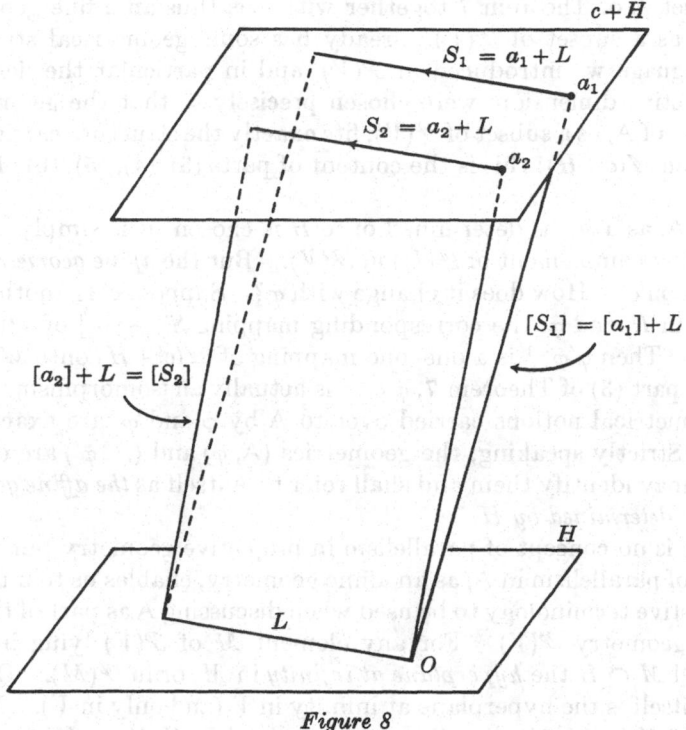

Figure 8

DEFINITION. Let S be a coset in a vector space V over a field F. A set A together with a one–one mapping φ of $\mathscr{A}(S)$ onto A is called an *affine geometry*.

There is an analogous generalization of projective geometry.

DEFINITION. Let V be a vector space over a field F. A set P together with a one–one mapping ψ of $\mathscr{P}(V)$ onto P is called a *projective geometry*.

The important thing to remember when one uses these generalized geometries is that any geometrical statement for A or P must always be interpreted in terms of the corresponding geometrical statement for $\mathscr{A}(S)$ or $\mathscr{P}(V)$ by using the link φ or ψ, respectively. Of course, if A or P, as sets, already have a geometrical structure in their own right, then we must be careful not to confuse this with the one carried over from $\mathscr{A}(S)$ or $\mathscr{P}(V)$ by the use of φ or ψ.

Strictly speaking, neither $\mathscr{A}(S)$ nor $\mathscr{P}(V)$ is a geometry in this generalized sense. But, naturally, they both become so if we use the identity mapping as our link.

The set A of Theorem 7 together with φ is thus an affine geometry. But A, as a subset of $\mathscr{P}(V)$, already has some geometrical structure. The language we introduced in $\mathscr{P}(V)$, and in particular the definition of projective dimension, were chosen precisely so that the geometrical structure of A, as a subset of $\mathscr{P}(V)$, fits exactly the structure carried over by φ from $\mathscr{A}(c+H)$: this is the content of parts (3), (4), (5), (6) of Theorem 7.

Now A, as a *set*, is determined once H is chosen: it is simply the set-theoretical complement of $\mathscr{P}(H)$ in $\mathscr{P}(V)$. But the *affine geometry* on A depends on φ. How does it change with φ? Suppose c' is another vector not in H and φ' the corresponding mapping $S' \to [S']$ of $\mathscr{A}(c'+H)$ onto A. Then $\varphi'\varphi^{-1}$ is a one–one mapping of $\mathscr{A}(c'+H)$ onto $\mathscr{A}(c+H)$ and, by part (3) of Theorem 7, $\varphi'\varphi^{-1}$ is actually an isomorphism. Thus the geometrical notions carried over to A by φ and φ' are exactly the same. Strictly speaking, the geometries (A, φ) and (A, φ') are distinct but we may identify them and shall refer to A itself as *the affine geometry in $\mathscr{P}(V)$ determined by H.*

There is no concept of parallelism in projective geometry but the existence of parallelism in A, as an affine geometry, enables us to introduce a suggestive terminology to be used when discussing A as part of the projective geometry $\mathscr{P}(V)$. For any element M of $\mathscr{P}(V)$ lying in A we shall call $M \cap H$ the *hyperplane at infinity* in M, or in $\mathscr{P}(M)$. Observe that H itself is the hyperplane at infinity in V (and only in V). For example, if M is a projective line not contained in H, then $M \cap H$ is the point at infinity in the line M; if N is a projective plane not contained in H, then $N \cap H$ is the line at infinity in the plane N.

Part (7) of Theorem 7 shows that the elements $S\varphi$, $T\varphi$ of A are parallel if, and only if, $S\varphi \cap H$ contains $T\varphi \cap H$ or vice versa; in other words, if, and only if, the hyperplane at infinity in $S\varphi$ contains, or is contained in, the hyperplane at infinity in $T\varphi$. Thus, in particular, two lines M, N are parallel in A if, and only if, as projective lines in $\mathscr{P}(V)$ they have the same point at infinity; a line and plane are parallel in A if, and only if, the point at infinity on the line lies on the line at infinity in the plane.

To sum up then, we may say that projective geometry is "little more" than affine geometry. Theorem 7 allows us to use either kind of geometry to study the other. For example, the reader may easily deduce the propositions of incidence in A or $\mathscr{A}(c+H)$ from those in $\mathscr{P}(V)$. This procedure is more illuminating than the use of Theorem 1 and indeed supersedes it; though, of course, Theorem 1 can now be deduced from Theorem 2 of Chapter I (cf. exercise 2 below).

The link between projective and affine geometry also enables us to draw pictures of projective configurations. Suppose that $\mathfrak{C}$ is a configuration in a projective plane $\mathscr{P}(V)$. We choose any projective line H and take the "cross-section" of $\mathfrak{C}$ with $c+H$ (where, of course, $c \notin H$). This cross-section yields a picture of an affine configuration which is actually the image of $\mathfrak{C}$ under the mapping φ^{-1} of Theorem 7. Different cross-sections may well give us pictures that look quite different, but they all represent the same projective configuration $\mathfrak{C}$. We would naturally choose a picture in which as little as possible of $\mathfrak{C}$ is lost off the paper.   For example, if $\mathfrak{C}$ is the projective triangle ABC, then the natural picture is Figure 9a, which is obtained by choosing H not to contain A, B, or C. The other two pictures in Figure 9 illustrate respectively the cases where H contains B (but not A or C) and when $H = BC$.

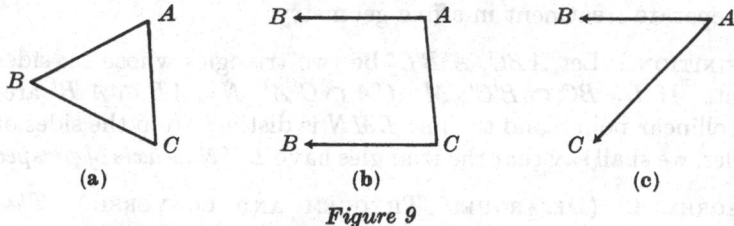

(a)   (b)   (c)

Figure 9

As a more elaborate example, suppose $\mathfrak{C}$ is the plane projective configuration shown in Figure 10a.   If we choose the line EG to be the line at infinity, then $\mathfrak{C}$ becomes the affine configuration pictured in Figure 10b (called a *parallelogram*).

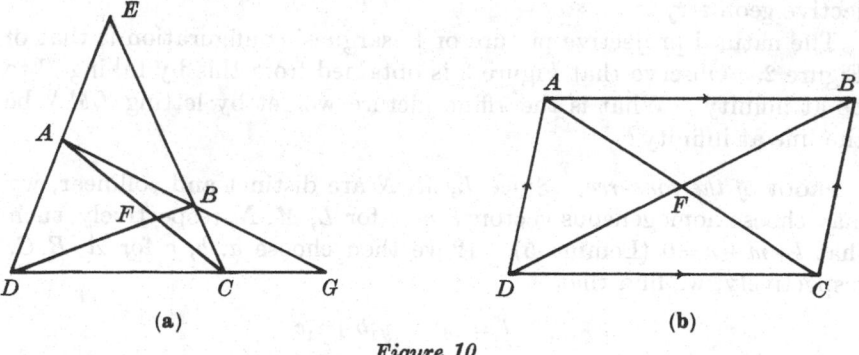

(a)   (b)

Figure 10

EXERCISES

1. If $\mathscr{P}(V)$ is a four-dimensional projective geometry, show that two planes of $\mathscr{P}(V)$ may intersect in a point.   Hence show that, if A is the affine geometry in $\mathscr{P}(V)$ determined by a hyperplane, two non-parallel planes in A may have their complete intersection "at infinity".

2. Let A be the affine geometry in $\mathscr{P}(V)$ determined by a hyperplane. Interpret Theorem 1 for A and prove it using only Theorem 7 and Theorem 2 of Chapter I.

3. Determine the number of lines in an affine plane over the field $\mathbf{F}_p$. (Embed the affine plane in a projective plane and use exercise 4 of § 2.5.)

2.7   The Fundamental Incidence Theorems of Projective Geometry

We shall now see that the theorems of Desargues and Pappus, as well as the Harmonic Construction, are essentially projective theorems: the statements are simpler than for the corresponding affine theorems, and moreover (by taking the hyperplane at infinity appropriately), each projective theorem leads to several affine theorems, each of which would require separate treatment in affine geometry.

DEFINITION.   Let ABC, $A'B'C'$ be two triangles whose six sides are distinct. If $L = BC \cap B'C'$, $M = CA \cap C'A'$, $N = AB \cap A'B'$ are distinct collinear points and the line LMN is distinct from the sides of the triangles, we shall say that the triangles have LMN as *axis of perspective*.

THEOREM 8. (DESARGUES' THEOREM AND CONVERSE.) *The triangles ABC, $A'B'C'$ have a center of perspective P if, and only if, they have an axis of perspective LMN.*

Note that the planes ABC, $A'B'C'$ may, or may not, be distinct. The direct part of the theorem is proved in essentially the same way as Theorem 3 (there being an obvious interpretation of Lemma 5 for projective geometry).

The natural projective picture of Desargues' configuration is that of Figure 2.   Observe that Figure 3 is obtained from this by taking P to be at infinity.   What is the affine picture we get by letting LMN be the line at infinity ?

PROOF *of the converse.*   Since L, M, N are distinct and collinear, we may choose homogeneous vectors l, m, n for L, M, N, respectively, such that $l + m + n = 0$ (Lemma 5).   If we then choose a, b, c for A, B, C, respectively, we find that

$$\begin{aligned} l &= & y_1 b + z_1 c \\ m &= x_2 a & + z_2 c \\ n &= x_3 a + y_3 b \end{aligned}$$

and so $(x_2 + x_3)a + (y_3 + y_1)b + (z_1 + z_2)c = 0$.

But a, b, c are linearly independent since A, B, C are not collinear, and hence $x_2 + x_3 = y_3 + y_1 = z_1 + z_2 = 0$.   None of these scalars can vanish because we have assumed that LMN is distinct from the sides of the triangle ABC. Thus we can find homogeneous vectors (again denoted by a, b, c) for A, B, C such that $l = b - c$, $m = c - a$, $n = a - b$.

Similarly we find a', b', c' for A', B', C', respectively, such that

$$l = b' - c', \quad m = c' - a', \quad n = a' - b',$$

and therefore $p = a - a' = b - b' = c - c' \neq 0$ is a homogeneous vector for a point P on AA', BB' and CC'. Finally, AA', BB' and CC' are distinct because we have assumed that the intersections $BC \cap B'C'$, etc., are points.

THEOREM 9. (PAPPUS' THEOREM.) *If two triads of points, A, B, C; A', B', C' are taken on two distinct coplanar lines and the intersections $BC' \cap B'C, CA' \cap C'A, AB' \cap A'B$ are points L, M, N, respectively, then L, M, N are collinear.*

The proof follows exactly the same pattern as that of Theorem 4. (Cf. Fig. 4.)

The above two theorems play a centrally important role in the synthetic foundations of projective geometry. Desargues' Theorem can always be deduced in a three dimensional projective space from the axioms of incidence 3.1–3.6 (p. 31), even if the planes of the triangles coincide (cf. exercise 3 below). On the other hand, there exist so called *non-Desarguesian projective planes* in which the axioms 2.1 and 2.2 (p. 31) are satisfied but Desargues' Theorem is *not* true. Clearly, such a non-Desarguesian plane cannot be embedded in any three dimensional projective space.

In all the work we have done so far, with the exception of one point in the proof of Pappus' Theorem, we have not used the commutative law of multiplication $xy = yx$ in the ground field F. A set F with rules of addition and multiplication satisfying all the axioms of a field, but with the possible exception of the commutative law of multiplication, is called a *division ring*. All our work on vector spaces and projective geometries can be extended to the case where F is a division ring, and all our theorems proved—*except* Pappus' Theorem. The method we used to establish Pappus' Theorem shows at once that it is true (for all configurations) in such a projective plane over a division ring if, and only if, the division ring is in fact a field.

There are two considerably deeper results which we mention without proof. The first is that Desargues' Theorem is true in a projective plane (defined synthetically by the propositions of incidence) if, and only if, this projective plane is a projective plane over a division ring (defined algebraically). The second is an algebraic theorem due to Wedderburn which states that a division ring with only a finite number of elements is a field. Combining these results gives a fascinating purely geometrical theorem: The truth of Desargues' Theorem in a *finite* projective plane implies the truth of Pappus' Theorem. (Consult [1], [6].)

THEOREM 10. (THE HARMONIC CONSTRUCTION.) *Given two distinct points A, B in a projective plane M and a point G on AB. We choose any point C in M but not on AB and also any point D on GC distinct from G and C. Then $E = AD \cap BC$, $F = BD \cap CA$ and $H = EF \cap AB$ are points and H is independent of the choice of C and D.*

The proof is almost identical to the one given for Theorem 5. (Cf. Fig. 5.)

DEFINITION. The point H is called the *harmonic conjugate* of G with respect to A, B. We also say that $(A, B; G, H)$ is a *harmonic range*.

For the harmonic construction in the projective plane it is no longer necessary to assume that the ground field has more than two elements. It is also clear that the exceptional cases where E, F or H failed to exist in the affine plane are not exceptions here. Observe that, on the affine line determined by choosing H as the point at infinity on AB, the mid-point of A, B—if it exists—is the harmonic conjugate of H with respect to A, B (cf. exercise 8, below).

The only subtle question remaining is the condition for a point G to coincide with its own harmonic conjugate. In fact $G = H$ if, and only if, $2xy = 0$ (exercise 3, § 2.4).

In an abstract field F we write 2 for the element $1 + 1$. Now it is possible that $2 = 0$ (e.g., in $\mathbf{F}_2$). A field in which $2 = 0$ is called a *field of characteristic 2*.

We see that there are just two cases:

(1) if F is a field of characteristic 2, then $G = H$ for all G on AB;

(2) if F is not a field of characteristic 2, then $G = H$ only for $G = A$ or $G = B$. (For then $2xy = 0$ implies $xy = 0$.)

DEFINITION. The configuration consisting of four coplanar points A, B, C, D, no three of which are collinear, and their six joins, is said to be a *(complete) quadrangle*. The points A, B, C, D are the *vertices* and their joins are the *sides* of the quadrangle. The intersections of the three pairs of opposite sides $AD \cap BC = E$, $BD \cap CA = F$, $CD \cap AB = G$ are called the *diagonal points*, and they are the vertices of the *diagonal point triangle* of the given quadrangle. (Cf. Fig. 10a.)

Our remarks above combine to give a proof of the following result.

THEOREM 11. (FANO'S THEOREM.) *If $\mathscr{P}(V)$ is a 2-dimensional projective geometry over a field F, then the diagonal points of every complete quadrangle in $\mathscr{P}(V)$ are collinear if, and only if, F has characteristic 2.*

We have seen above that the harmonic construction is valid in a projective geometry over the field $\mathbf{F}_2$. If the planes ABC, $A'B'C'$ in Desargues' Theorem (Theorem 8) are assumed to be distinct then there is no need to restrict the ground field; but if we try to draw the configuration of Desargues' Theorem in a projective plane over $\mathbf{F}_2$, then by Fano's Theorem one at least of the triads A, B, C; A', B', C' must be collinear. Thus for the plane form of Desargues' Theorem we must assume that the ground field contains at least three elements.

EXERCISES

1. If two tetrahedra $A_0A_1A_2A_3$, $B_0B_1B_2B_3$ are in perspective from a point P then the six points of intersection $A_iA_j \cap B_iB_j = P_{ij}$, $0 \le i < j \le 3$, are coplanar.

2. Deduce four different affine theorems from Theorem 8.

3. Given two *coplanar* triangles ABC, $A'B'C'$ in perspective from a point P. Take any line through P not in the plane of the triangles and any points Q, R on this line so that P, Q, R are distinct. Find a triangle $A''B''C''$ in perspective with ABC from Q and in perspective with $A'B'C'$ from R. Hence deduce Desargues' Theorem for the given coplanar triangles by using Desargues' Theorem for non-coplanar triangles (cf. § 2.5, exercise 6).

4. Let $A_iB_iC_i$, $i = 1, 2, 3$, be three triangles, any two of which are in perspective, the center of perspective of $A_iB_iC_i$ and $A_jB_jC_j$ being P_{ij}. Assume P_{23}, P_{31}, P_{12} are distinct collinear points and that $A_1A_2A_3$, $B_1B_2B_3$, $C_1C_2C_3$ are triangles. Prove that $A_iB_iC_i$, $i = 1, 2, 3$ have a common axis of perspective L, and that any two of $A_1A_2A_3$, $B_1B_2B_3$, $C_1C_2C_3$ are in perspective, the three centers of perspective lying on L.

5. By considering the triangles $B'BN$, $C'CM$ show that the converse of Desargues' Theorem can be deduced from the direct theorem in the two-dimensional case.

6. Show that Pappus' Theorem can be deduced from Desargues' Theorem in the case where the two triads are in perspective.

7. Using the notations of Theorem 9 (Pappus' Theorem) and assuming that A, B, C, A', B', C' are distinct from $P = ABC \cap A'B'C'$, prove that the line LMN contains P if, and only if, AA', BB', CC' are concurrent.

8. Let A, B be distinct points on an affine line L. By embedding L in a projective line, verify that there exists a point on L having no harmonic conjugate with respect to A, B if, and only if, the ground field is not of characteristic 2.

9. Draw a complete quadrangle and join the pairs of diagonal points. Mark in all the harmonic ranges in your diagram.

10. Prove that the diagonals of a parallelogram bisect each other. (Cf. Fig. 10b.)

CHAPTER III

Isomorphisms

3.1 Affinities

If S, S' are cosets in vector spaces V, V' over fields F, F', respectively, we shall often consider mappings f of S into S' with the following property: for any coset T in S, the image set Tf is a coset in S'. In such a case the mapping

$$T \to Tf$$

of $\mathscr{A}(S)$ into $\mathscr{A}(S')$ induced by f will be denoted by $\mathscr{A}(f)$.

Clearly, if f, g are two such mappings for which the product fg is defined then

$$\mathscr{A}(fg) = \mathscr{A}(f)\mathscr{A}(g).$$

It is important to note that f and $\mathscr{A}(f)$ are logically distinct: the former acts on vectors and the latter acts on cosets. On the other hand we see that f may be regained from $\mathscr{A}(f)$ merely by restricting $\mathscr{A}(f)$ to the cosets consisting of one vector: explicitly,

$$\{v\}\mathscr{A}(f) = \{v'\} \quad \text{if, and only if,} \quad vf = v'.$$

Occasionally we shall find it convenient to write $v\,\mathscr{A}(f)$ instead of $\{v\}\mathscr{A}(f)$.

If a is a vector in V then the *translation*

$$t_a : v \to v + a$$

induces a mapping

$$\mathscr{A}(t_a) : T \to T + a,$$

where T is a coset in V. We shall refer to $\mathscr{A}(t_a)$ as a *translation* in $\mathscr{A}(V)$. If S is a coset in V, the restrictions of t_a and $\mathscr{A}(t_a)$ to S and $\mathscr{A}(S)$, respectively, will still be denoted by the same letters.

DEFINITION. If $S = a + M$, $S' = a' + M'$ are cosets in vector spaces V, V' over the same field F and if g is an isomorphism of the subspace M onto the subspace M', then a mapping of the form

$$\mathscr{A}(t_{-a}\, g\, t_{a'})$$

will be called an *affinity* of $\mathscr{A}(S)$ onto $\mathscr{A}(S')$.

42

It is obvious that an affinity is a particular kind of isomorphism—in fact, the isomorphism constructed in proving Theorem 2 of Chapter II (p. 21) was an affinity. We shall prove later that any isomorphism of affine geometries over different fields is as near to an affinity as one could possibly expect (see Theorem 5). For affine geometries over the same field, affinities are the most natural isomorphisms to study.

Let us denote $t_{-a}\, g\, t_{a'}$ by h. If s is any vector in S then

$$\begin{aligned}
h &= t_{-s}\, t_{s-a}\, g\, t_{a'} \\
&= t_{-s}\, g\, t_{(s-a)g+a'}, \quad \text{by the linearity of } g, \\
&= t_{-s}\, g\, t_{s'},
\end{aligned}$$

where $s' = (s-a)g + a' = sh$. It follows that

$$g = t_s\, h\, t_{-sh}.$$

But s was an arbitrary element of S and thus g is *uniquely determined* by h.

When S is a subspace it is natural to choose $s = 0$ and to express h uniquely in the form $g\, t_{s'}$ where $s' = 0h$.

It is easy to verify that the product of two affinities, where defined, is again an affinity. For if $\alpha = \mathscr{A}\, (t_{-s}\, g\, t_{s'})$ is an affinity of $\mathscr{A}(S)$ onto $\mathscr{A}(S')$, then an affinity of $\mathscr{A}(S')$ onto $\mathscr{A}(S'')$ can be written in the form $\alpha' = \mathscr{A}(t_{-s'}\, g'\, t_{s''})$ so that the product $\alpha\alpha'$ is $\mathscr{A}(t_{-s}\, g\, g'\, t_{s''})$.

In order to introduce coordinates in an affine geometry $\mathscr{A}(S)$ it is natural to attempt to use a basis in the underlying vector space V. But even if the geometry $\mathscr{A}(S)$ contains the point O (corresponding to the zero vector), there is no geometrical feature to distinguish this from any other point and so there is no purely geometrical notion which exactly corresponds to that of basis.

DEFINITION. Let A be an affine geometry of dimension $n \geq 1$ and let $P_1, \ldots, P_n,\, Q$ be $n+1$ points of A whose join has the maximum dimension n. Then the ordered $(n+1)$-tuple $(P_1, \ldots, P_n,\, Q)$ is called a *frame of reference* for A, with *origin Q*.

We recall from Chapter I (p. 13) that the space F^n has a *standard basis* $\{e_1, \ldots, e_n\}$ where e_i is the ordered n-tuple $(0, \ldots, 0, 1, 0, \ldots, 0)$ with i-term equal to 1 and all the other terms equal to 0 $(i = 1, \ldots, n)$. If we denote the points $\{e_i\}$ by E_i and the point $\{(0, \ldots, 0)\}$ by O then $(E_1, \ldots, E_n,\, O)$ is the *standard frame of reference* for $\mathscr{A}(F^n)$ with origin O.

The following result is a stronger form of Theorem 2 of Chapter II.

THEOREM 1. *If $(P_1, \ldots, P_n,\, Q)$ and $(P_1', \ldots, P_n',\, Q')$ are frames of reference for affine geometries A, A', respectively, defined over the same field, then there is a unique affinity α of A onto A' such that $P_i\alpha = P_i'$ for $i = 1, \ldots, n$ and $Q\alpha = Q'$.*

PROOF. We may suppose that $A = \mathscr{A}(a + M)$ and $A' = \mathscr{A}(a' + M')$ where $a + M$, $a' + M'$ are cosets in vector spaces V, V', respectively, over the same field. Let the points $P_1, \ldots, P_n, Q, P_1', \ldots, P_n', Q'$ have vectors $p_1, \ldots, p_n, q, p_1', \ldots, p_n', q'$, respectively.

Any affinity α which satisfies $Q\alpha = Q'$ is expressible in the form

$$\alpha = \mathscr{A}(t_{-q}\, g\, t_{q'})$$

where g is an isomorphism of M onto M'. If also $P_i\alpha = P_i'$ for $i = 1, \ldots, n$ then g takes $p_i - q$ into $p_i' - q'$ for $i = 1, \ldots, n$. By the definition of frames of reference, the vectors $p_i - q$ $(i = 1, \ldots, n)$ form a basis of M and the vectors $p_i' - q'$ $(i = 1, \ldots, n)$ form a basis of M' (cf. exercise 4 of § 2.2, p. 20). Thus by Theorem 3 of Chapter I (p. 12) there is one and only one such isomorphism g and so there is one and only one such affinity α.

DEFINITION. If A is an affine geometry of dimension n over a field F then an affinity α of A onto $\mathscr{A}(F^n)$ is an *affine coordinate system* for A.

It is clear from Theorem 1 that there is a one-one correspondence between the frames of reference $(P_1, \ldots, P_n, Q)$ for A and the affine coordinate systems α for A, given by

$$P_i\alpha = E_i \quad (i = 1, \ldots, n), \qquad Q\alpha = O.$$

This affinity α takes the point $X = \{q + \sum x_i(p_i - q)\}$ of A to the point $\{(x_1, \ldots, x_n)\}$ of $\mathscr{A}(F^n)$. We shall say that $(x_1, \ldots, x_n)$ is the *coordinate row* of X with respect to the given frame of reference (or, with respect to the given affine coordinate system).

EXERCISES

1. If α is an affinity of $\mathscr{A}(S)$ onto $\mathscr{A}(S')$ and T is a coset contained in S, prove that the restriction of α to $\mathscr{A}(T)$ is also an affinity.

2. Prove that the inverse of the affinity $\mathscr{A}(t_{-a}\, g\, t_{a'})$ is the affinity $\mathscr{A}(t_{-a'}\, g^{-1}\, t_a)$.

3. Given $\alpha = \mathscr{A}(t_{-a}\, g\, t_{a'})$: $\mathscr{A}(a + M) \to \mathscr{A}(a + M)$, prove that the following two properties are equivalent:
   (i) $S\alpha$ is parallel to S for every coset S contained in $a + M$;
   (ii) $Ng = N$ for every subspace N of M.
   (An affinity with either of these properties is called a *dilatation*. Cf. exercises 9 of § 3.2 and 1 of § 3.4, where these mappings are discussed further.)

3.2 Projectivities

Let V and V' be two vector spaces over the same field. If f is a mapping of V into V' such that, for every subspace M of V, the image set Mf is always a subspace of V', then f induces a mapping

$$M \to Mf$$

of $\mathscr{P}(V)$ into $\mathscr{P}(V')$. We shall denote this by $\mathscr{P}(f)$ and stress that, as in the analogous affine case, f acts on vectors while $\mathscr{P}(f)$ acts on subspaces.

DEFINITION. If f is an isomorphism of V onto V' then $\mathscr{P}(f)$ is called a *projectivity* of $\mathscr{P}(V)$ onto $\mathscr{P}(V')$. In particular, if $V' = V$, then $\mathscr{P}(f)$ is called a *collineation* of $\mathscr{P}(V)$.

It is clear that if f and g are two mappings for which $\mathscr{P}(f)$, $\mathscr{P}(g)$ and the product fg are defined, then $\mathscr{P}(fg)$ is also defined and

$$\mathscr{P}(fg) = \mathscr{P}(f)\mathscr{P}(g).$$

Further, if $\mathscr{P}(f)$ and $\mathscr{P}(g)$ are projectivities, so is $\mathscr{P}(f)\mathscr{P}(g)$.

A projectivity is clearly an isomorphism of the projective geometries. We shall find in § 3.5 that the most general isomorphism of $\mathscr{P}(V)$ onto $\mathscr{P}(V')$, when the ground fields are distinct, is as close to being a projectivity as is possible (see Theorem 6).

If f is an isomorphism of V onto V' and z any non-zero scalar, then $g = zf$ is the isomorphism defined by the rule $vg = z(vf)$ for all v in V.

PROPOSITION 1. *Two isomorphisms f, g of V onto V' induce the same projectivity $\mathscr{P}(f) = \mathscr{P}(g)$ if, and only if, there exists a non-zero scalar z such that $g = zf$.*

PROOF. If $g = zf$ then $Mf = Mg$ for any subspace M of V and so $\mathscr{P}(f) = \mathscr{P}(g)$.

Conversely, suppose that $\mathscr{P}(f) = \mathscr{P}(g)$; then $ag = z_a(af)$ where z_a is a non-zero scalar possibly depending on the vector a.

Assume first that $V = [a]$. Then

$$\begin{aligned}
(xa)g &= x(ag) \\
&= x(z_a(af)) \\
&= (xz_a)(af) \\
&= (z_ax)(af) \\
&= z_a(xa)f, \quad \text{for any } x \text{ in } F.
\end{aligned}$$

In other words $z_{xa} = z_a$ for any non-zero x in F.

Assume that dim $V \geq 2$. If a, b are linearly independent vectors in V then af, bf are linearly independent in V', so the equation

$$(a+b)g = ag + bg,$$

which is equivalent to

$$z_{a+b}(a+b)f = z_a(af) + z_b(bf),$$

yields

$$z_a = z_b = z_{a+b}.$$

If a and b are linearly dependent we may use the above argument for the one-dimensional case to show that $z_a = z_b$. Alternatively, and this argument has an application later (pp. 57, 122), we may find a vector c in V not linearly dependent on a or b, and deduce that $z_a = z_c$, $z_b = z_c$ whence $z_a = z_b$. This establishes the result.

We turn now to the problem of introducing coordinates.

DEFINITION.   If P is a projective geometry of dimension n over the field F then a projectivity π of P onto $\mathscr{P}(F^{n+1})$ is a *projective (or homogeneous) coordinate system* for P.

We look for a frame of reference in P which will determine π uniquely. Let us write e_i for the $(n+1)$-tuple $(0, \ldots, 1, \ldots, 0)$ with i-term equal to 1 and all other terms equal to 0 $(i = 0, \ldots, n)$. (Note that the i-term is in the $(i+1)$ position here, since we number from 0 to n.) If $A_i \pi = [e_i]$ for $i = 0, \ldots, n$ then the (ordered) $(n+1)$-tuple $(A_0, \ldots, A_n)$ is often called the *(ordered) simplex of reference* for the given coordinate system π.

Suppose that the simplex $(A_0, \ldots, A_n)$ is given. We attempt to reconstruct π as follows: Let $a_0, \ldots, a_n$ be any homogeneous vectors for $A_0, \ldots, A_n$, respectively (i.e., $[a_i] = A_i$ for $i = 0, \ldots, n$). Then there is a unique isomorphism f of V onto F^{n+1} taking a_i to e_i for each i, and the projectivity $\pi = \mathscr{P}(f)$ certainly takes A_i to $[e_i]$ for each i. Unfortunately there is an ambiguity in the choice of each a_i (except in the very special case where the ground field F has only two elements 0, 1), because each homogeneous vector may be multiplied by an arbitrary non-zero scalar in F. When $n = 0$ the ambiguity is "absorbed" when we take $\mathscr{P}(f)$ (cf. Proposition 1). When $n \geq 1$, however, we need a further point to "stiffen up" our simplex of reference.

DEFINITION.   Let P be a projective geometry of dimension $n \geq 1$ and suppose that $(A_0, \ldots, A_n, U)$ is an ordered $(n+2)$-tuple of points in P such that the sum (join) of every subset consisting of $(n+1)$ of these points is of the maximum dimension n. Then $(A_0, \ldots, A_n, U)$ is called a *frame of reference* for P with *unit point* U and *simplex* $(A_0, \ldots, A_n)$.

For example, in a 2-dimensional projective geometry any triangle may be taken as triangle of reference, and any point, not on any of the sides of the triangle, as unit point. In a three-dimensional projective geometry we may take any tetrahedron as tetrahedron of reference, and any point, not on any of the faces, as unit point.

The ordered basis $(a_0, \ldots, a_n)$ of V is said to *determine* the frame of reference $(A_0, \ldots, A_n, U)$ for $\mathscr{P}(V)$ if $A_i = [a_i]$ for $i = 0, \ldots, n$ and

$U = [a_0 + \cdots + a_n]$. The standard basis of F^{n+1} determines the *standard frame of reference* for $\mathscr{P}(F^{n+1})$, viz., $(E_0, \ldots, E_n, E)$ where $E_i = [e_i]$ for $i = 0, \ldots, n$ and $E = [(1, 1, \ldots, 1)]$.

LEMMA 1. *Every frame of reference for $\mathscr{P}(V)$ is determined by at least one ordered basis of V. Two ordered bases $(a_0, \ldots, a_n)$, $(b_0, \ldots, b_n)$ of V determine the same frame of reference if, and only if, there exists a nonzero scalar z such that $b_i = za_i$ for $i = 0, \ldots, n$.*

PROOF. Consider the frame of reference $(A_0, \ldots, A_n, U)$. Let u be a homogeneous vector for U (i.e., $[u] = U$) and let $c_0, \ldots, c_n$ be any homogeneous vectors for $A_0, \ldots, A_n$, respectively. Then $(c_0, \ldots, c_n)$ is an ordered basis of V and $u = x_0 c_0 + \cdots + x_n c_n$ where $x_i \neq 0$ for $i = 0, \ldots, n$. If we put $a_i = x_i c_i$ then $[a_i] = A_i$ $(i = 0, \ldots, n)$ and $u = a_0 + \cdots + a_n$.

If $(a_0, \ldots, a_n)$, $(b_0, \ldots, b_n)$ determine the same frame of reference, then $b_i = z_i a_i$ $(i = 0, \ldots, n)$ and $b_0 + \cdots + b_n = z(a_0 + \cdots + a_n)$. Hence $z_0 a_0 + \cdots + z_n a_n = za_0 + \cdots + za_n$ from which it follows, by the linear independence of $a_0, \ldots, a_n$, that $z_0 = \cdots = z_n = z$.

THEOREM 2. *If $(A_0, \ldots, A_n, U)$, $(A_0', \ldots, A_n', U')$ are frames of reference for projective geometries P, P' over the same field, then there exists a unique projectivity π of P onto P' such that*

$$A_i \pi = A_i' \quad \text{for } i = 0, \ldots, n \quad \text{and} \quad U\pi = U'.$$

PROOF. We consider geometries $\mathsf{P} = \mathscr{P}(V)$, $\mathsf{P}' = \mathscr{P}(V')$. By Lemma 1 there exist ordered bases $(a_0, \ldots, a_n)$, $(a_0', \ldots, a_n')$ determining the given frames of reference. By Theorem 3 of Chapter I (p. 12) there is an isomorphism f of V onto V' such that $a_i f = a_i'$ for $i = 0, \ldots, n$ and consequently $(a_0 + \cdots + a_n)f = a_0' + \cdots + a_n'$. The projectivity $\mathscr{P}(f)$ clearly satisfies the requirements of the theorem.

Now suppose that $\mathscr{P}(g)$ is another such projectivity. Then the ordered bases $(a_0 f, \ldots, a_n f)$, $(a_0 g, \ldots, a_n g)$ of V' determine the same frame of reference $(A_0', \ldots, A_n', U')$. Hence, by Lemma 1, $a_i g = za_i f$ for each i. Therefore $g = zf$ and so $\mathscr{P}(f) = \mathscr{P}(g)$ by Proposition 1.

By Theorem 2 there is a one-one correspondence between frames of reference $(A_0, \ldots, A_n, U)$ for P and projective coordinate systems π for P given by

$$A_i \pi = E_i \quad \text{for } i = 0, \ldots, n \quad \text{and} \quad U\pi = E.$$

If P is a point in the geometry P and $P\pi = [(x_0, \ldots, x_n)]$, then we say that $(x_0, \ldots, x_n)$ is a *projective* (or *homogeneous*) *coordinate row* of the point P with respect to π (or with respect to the frame of reference corresponding to π).

EXERCISES

1. If $\mathscr{P}(f)$ is a projectivity, prove that $\mathscr{P}(f)^{-1} = \mathscr{P}(f^{-1})$.

2. When the ground field is **R**, we define the *centroid* of the affine points $a_0, \ldots, a_n$ to be the point $\dfrac{1}{n+1}(a_0 + \cdots + a_n)$. Show that Theorem 2 yields the following affine theorem:

   Given a real affine geometry of dimension $(n+1)$ and $(n+1)$ linearly independent lines $A_0, \ldots, A_n$ through the point O together with a further line U through O which is not linearly dependent on any n of the A_i's. Then there exist hyperplanes cutting $A_0, \ldots, A_n$, U in $(n+1)$ distinct points and their centroid, respectively; and any two such hyperplanes are parallel. (The cases $n = 1, 2$ are easily visualized.)

3. Let P be a projective plane and π a collineation having a line L of fixed points (i.e., $X\pi = X$ for all points X on L). If π is not the identity mapping, prove that there exists a point A (possibly on L) such that A, P and $P\pi$ are collinear for every point P in P.

   (Such a collineation π is called a *central collineation*. The line L is the *axis* and the point A is the *center*. Cf. also § 3.7.)

4. Let P be a projective plane and suppose that we are given a line L, distinct points P, P' not on L and a point A on PP' but distinct from P and P'. Prove that there exists one, and only one, central collineation π of P with axis L, center A and such that $P\pi = P'$. (For the definitions see immediately above.)

5. Let ABC, $A'B'C'$ be two triangles in perspective from a point P. Let $L = BC \cap B'C'$, $M = CA \cap C'A'$, $N = AB \cap A'B'$. By the previous exercise there exists a central collineation π with axis LM, center P and satisfying $C\pi = C'$. Deduce that L, M, N are collinear.

   (This exercise shows that the truth of Desargues' Theorem in a projective plane is closely linked to the existence of sufficiently many central collineations. This fact is of importance in the study of non-Desarguesian projective planes. Cf. remarks on page 39. For further information the reader should consult [3], [6].)

6. If L, L' are distinct coplanar lines, A is a point on neither line and $P \rightarrow P'$ is the mapping of the points of L onto those of L' defined by $P' = AP \cap L'$, prove that there exist central collineations π such that $P\pi = P'$ for all P on L.

   (The mapping $P \rightarrow P'$ yields, in an obvious way, an isomorphism of $\mathscr{P}(L)$ onto $\mathscr{P}(L')$. Such an isomorphism is called a *perspectivity* and A is its *center*.)

7. Let π be a collineation of a projective plane P which takes a line L in P into a distinct line L'. Use the uniqueness part of Theorem 2 to show that the induced projectivity of $\mathscr{P}(L)$ onto $\mathscr{P}(L')$ is a perspectivity if (and only if) the point $L \cap L'$ is left fixed by π.

8. Show that a projectivity of the line $\mathscr{P}(L)$ onto the distinct coplanar line $\mathscr{P}(L')$ can always be expressed as a product of two perspectivities.

Prove, further, that a collineation on $\mathscr{P}(L)$ can always be written as a product of three perspectivities.

9. Using exercise 3 of § 3.1, prove that an affinity α of $\mathscr{A}(a+M)$ onto $\mathscr{A}(a+M)$ is a dilatation if, and only if, the uniquely determined vector space isomorphism associated with α is $x1_M$, for some non-zero scalar x. Show, further, that a dilatation (other than the identity) is a translation if, and only if, it does not fix any point.   (Cf. also exercise 1 of § 3.4.)

10. Let V and V' be non-zero vector spaces over a division ring D (cf. p. 39) and let f be an isomorphism of V onto V'.   If z is a non-zero element of D show that the mapping

$$zf: v \rightarrow z(vf)$$

is an isomorphism if, and only if, $zx = xz$ for every x in D.

   Show that the *uniqueness* part of Theorem 2 cannot be extended to geometries P, P' over a division ring D which is not commutative.   (We remarked on page 39 that the truth of Pappus' Theorem in a projective plane over D is sufficient to ensure that D is commutative.   This exercise gives another such criterion.)

11. Let z, x be elements of a division ring D such that $zx \neq xz$ and let M, N be the subspaces of D^3 spanned by $\{(1, 0, 0), (0, 1, 0)\}$, $\{(1, 0, 0), (0, 0, 1)\}$, respectively.   If g is the isomorphism of M onto N defined by $(1, 0, 0)g = (z, 0, 0)$ and $(0, 1, 0)g = (0, 0, z)$, show that the projectivity $\mathscr{P}(g)$ leaves $[(1, 0, 0)]$ fixed but is *not* a perspectivity (cf. exercise 7).

3.3   Linear Equations

One of the great advantages of coordinates is that they allow the use of *equations*. The following theorems show the relationship which exists between *linear* equations and the geometrical objects we have been studying.

THEOREM 3.   *The set of all solutions* $(x_1, \ldots, x_n)$ *in* F^n *of the linear equations*

$$\sum_{j=1}^{n} a_{ij}X_j = b_i \quad (i = 1, \ldots, m) \tag{1}$$

with all a_{ij}, b_i *in* F, *is either empty or is a coset of dimension* $n-r$ *in* F^n, *where* r *is the maximum number of linearly independent rows* $(a_{i1}, \ldots, a_{in})$, $i = 1, \ldots, m$.

If $c = (c_1, \ldots, c_n)$ is any solution of the equations (1) then (by the linearity) $x = (x_1, \ldots, x_n)$ is a solution of the linear equations

$$\sum_{j=1}^{n} a_{ij}X_j = 0 \quad (i = 1, \ldots, m) \tag{2}$$

if, and only if, $x + c$ is a solution of the equations (1).   The proof of Theorem 3 reduces therefore to the proof of

THEOREM 4. *The set of all solutions $(x_1, \ldots, x_n)$ in F^n of the (homogeneous) linear equations* (2) *is a subspace of F^n of dimension $(n-r)$ where r is the maximum number of linearly independent rows $(a_{i1}, \ldots, a_{in})$, $i = 1, \ldots, m$.*

It is convenient to introduce the subspace of F^n spanned by the rows $(a_{i1}, \ldots, a_{in})$, $(i = 1, \ldots, m)$. We call this subspace the *coefficient space* of the system of equations (1) or (2). The number r appearing in the statements of Theorems 3 and 4 is the dimension of the coefficient space.

PROOF OF THEOREM 4. If $r = 0$ then the theorem is true since then every element of F^n is a solution. We suppose therefore that one of the coefficients a_{ij} is non-zero. By suitably renumbering the equations and variables, if necessary, we may assume that $a_{11} \neq 0$. We shall write L_i for $\sum_{j=1}^{n} a_{ij} X_j$, $i = 1, \ldots, m$. If now $L_1' = L_1$, but

$$L_i' = L_i - (a_{i1}/a_{11})L_1$$

for $i = 2, \ldots, m$, then the system $L_i' = 0$ $(i = 1, \ldots, m)$ has exactly the same set of solutions and the same coefficient space as the system $L_i = 0$ $(i = 1, \ldots, m)$. This follows at once from the fact that every L_i' is a linear combination of the L_j, and every L_i is a linear combination of the L_j'.

We reduce the equations step by step in this way (with suitable renumberings, if necessary) until a "truncated triangle" of equations is obtained, say

$$
\begin{aligned}
b_{11}X_1 + \cdots &\qquad\qquad + b_{1n}X_n = 0, \\
b_{22}X_2 + \cdots &\qquad\qquad + b_{2n}X_n = 0, \\
&\cdots \\
+ b_{ss}X_s &+ \cdots + b_{sn}X_n = 0,
\end{aligned}
\tag{3}
$$

where $b_{11}, b_{22}, \ldots, b_{ss}$ are non-zero. Since the dimension of the coefficient space is unaltered at each step, we see that $s = r$. Equations (3) are now solved by giving $X_{r+1}, \ldots, X_n$ arbitrary values in F and solving uniquely for $X_r, X_{r-1}, \ldots, X_1$ in turn, starting with the last equation and working upwards.

If $r = n$, the only solution is $(0, 0, \ldots, 0)$; but if $r < n$, the solutions are of the form

$$x_{r+1}a_{r+1} + \cdots + x_n a_n$$

where, for each $j = r+1, \ldots, n$, a_j is a row of which the last $(n-r)$ terms are $(0, \ldots, 0, 1, 0, \ldots, 0)$ with 1 as the $(j-r)$-term. These a_j are therefore linearly independent and the result follows.

Our method of proving these theorems is less satisfactory, from the point of view of elegance and speed, than others to be given later, but it does have the advantage of illustrating the most effective practical method of solving linear equations.

Theorem 3 has the following partial converse.

PROPOSITION 2. *Every coset in F^n is the set of all solutions of a suitable system of equations* (1).

PROOF. By the reduction already given of Theorem 3 to Theorem 4 we need only prove our proposition when the given coset is actually a subspace M of F^n.

We define $M^\perp$ to be the set of all $(y_1, \ldots, y_n)$ in F^n satisfying

$$\sum_{i=1}^n x_i y_i = 0$$

for all $(x_1, \ldots, x_n)$ in M. If we interpret M as a coefficient space we see that $M^\perp$ is a subspace of F^n of dimension $n - \dim M$.

The subspace $M^{\perp\perp}$ consisting of all $(z_1, \ldots, z_n)$ in F^n satisfying

$$\sum_{i=1}^n y_i z_i = 0$$

for all $(y_1, \ldots, y_n)$ in $M^\perp$ clearly contains M. But $M^{\perp\perp}$ has dimension $n - (n - \dim M) = \dim M$ and so $M^{\perp\perp} = M$ (by Theorem 2(1) of Chapter I (p. 11)).

If the rows $(a_{i1}, \ldots, a_{in})$, $i = 1, \ldots, m$, span $M^\perp$ then M is the solution space of the equations (2).

There is a natural interpretation of these results for an affine or a projective geometry for which a coordinate system is given.

If A is an affine geometry with a coordinate system α and S an element of A, then by Proposition 2, $S\alpha$ is the set of all solutions of a suitable system of linear equations (1). It is usual to say that the equations (1) are a *set of equations for S in the given coordinate system*. There is a similar terminology for projective geometries.

EXERCISES

1. A single equation $a_1 X_1 + \cdots + a_n X_n = b$, with not all $a_i = 0$, defines a hyperplane in $\mathscr{A}(F^n)$ which is also a hyperplane of $\mathscr{P}(F^n)$ if, and only if, $b = 0$.

2. Prove Desargues' Theorem by taking ABC as triangle of reference and P as unit point.

3. Prove Pappus' Theorem by taking PAA' as triangle of reference and N as unit point.

4. Prove exercise 1 of § 2.7 by taking $A_0 A_1 A_2 A_3$ as tetrahedron of reference and P as unit point.

5. Reinterpret the proof of Theorem 10 of Chapter II (the Harmonic Construction) by taking ABC as triangle of reference. When G is distinct from A and B (and the ground field is not a field of characteristic 2) then EFG may be taken as triangle of reference and D as unit point. Show that then A, B, C have coordinate rows $(-1, 1, 1)$, $(1, -1, 1)$, $(1, 1, -1)$, respectively.

6. In the notation of the Embedding Theorem of Chapter II (p. 32), choose a basis $\{a_0, \ldots, a_n\}$ of V such that $a_0 = c$ and $[a_1, \ldots, a_n] = H$. What are the equations of H and $c + H$? Show that the affine point with coordinates $(1, x_1, \ldots, x_n)$ is mapped by φ to the projective point with homogeneous coordinates $(1, x_1, \ldots, x_n)$, and that the coset in $c + H$ defined by the equations

$$\sum_{j=1}^{n} a_{ij}X_j = b_i, \quad i = 1, \ldots, m,$$

is mapped by φ onto the subspace with equations

$$\sum_{j=1}^{n} a_{ij}X_j = b_iX_0, \quad i = 1, \ldots, m.$$

7. Let π be a projective coordinate system of $\mathscr{P}(V)$ and H the hyperplane with equation $X_0 = 0$. If A is the affine geometry determined by H and P is a point in A with projective coordinate row $(x_0, \ldots, x_n)$, prove that

$$P \rightarrow (x_1/x_0, \ldots, x_n/x_0)$$

is an affine coordinate system of A. Identify the origin of the corresponding affine frame of reference.

3.4 Affine and Projective Isomorphisms

In view of the fundamental theorem about the embedding of affine geometries in projective geometries it is natural to expect that affine and projective isomorphisms are closely related.

Suppose that $\mathscr{P}(V)$, $\mathscr{P}(V')$ are projective geometries, not necessarily over the same field, and that π is an isomorphism of $\mathscr{P}(V)$ onto $\mathscr{P}(V')$. We may choose any hyperplane H in $\mathscr{P}(V)$ and the corresponding hyperplane $H' = H\pi$ in $\mathscr{P}(V')$ as hyperplanes at infinity and so obtain two affine geometries A, A', respectively (p. 36). The restriction α of π to A is immediately seen to be an isomorphism of A onto A'.

Conversely, a given isomorphism α of A onto A' may be extended in one and only one way to an isomorphism π of $\mathscr{P}(V)$ onto $\mathscr{P}(V')$, as follows. We define $P\pi = P\alpha$ for any P in A. Since H is the only hyperplane in $\mathscr{P}(V)$ not contained in A, and similarly for H', we must set $H\pi = H'$. If M is any subspace of H then all the elements P of A which have M as hyperplane at infinity (i.e., $P \cap H = M$) are parallel and have the same dimension. (Such elements P certainly exist, for if $c \notin H$ then $P = [c] + M$ has M as hyperplane at infinity.) The corresponding elements $P\alpha$ of A′ are again all parallel and have the same dimension; as such they all have the same hyperplane at infinity M'. We are thus forced to define $M\pi = M'$. It only remains to show that the one–one mapping π so defined, and its inverse π^{-1}, preserve inclusion.

Since π extends α it is clear that π preserves inclusions among the elements of A. Suppose that M, N are subspaces of H and that P, Q respectively have M, N as hyperplanes at infinity. Then $M \subset N$ if, and only if, P and Q are parallel and dim $P \le$ dim Q. By our definition, $M\pi \subset N\pi$ if, and only if, $P\alpha$ and $Q\alpha$ are parallel and dim $P\alpha \le$ dim $Q\alpha$. Since α is an isomorphism it follows that $M \subset N$ if, and only if, $M\pi \subset N\pi$.

If M is a subspace of H and Q is an element of A with hyperplane at infinity $N = Q \cap H$, then $M \subset Q$ if, and only if, $M \subset N$. From this it now follows that $M \subset Q$ if, and only if, $M\pi \subset Q\pi$.

Suppose now that $\mathscr{P}(V)$, $\mathscr{P}(V')$ are defined over the same field F and that π is a projectivity. In other words $\pi = \mathscr{P}(f)$ where f is an isomorphism of V onto V' taking H onto H'. The geometries (A, φ), (A′, φ') are linked to geometries $\mathscr{A}(c+H)$, $\mathscr{A}(c'+H')$ where c, c' are arbitrary elements of V, V' not in H, H', respectively. By multiplying f by a suitable non-zero scalar, if necessary, we may assume that cf lies in $c'+H'$. We may then assume without loss of generality that $c' = cf$. We wish to prove that the affine isomorphism α of A onto A′ determined by π (as explained above) is here an affinity. This amounts to proving that the induced mapping $\varphi\alpha\varphi'^{-1}$ of $\mathscr{A}(c+H)$ onto $\mathscr{A}(c'+H')$ is an affinity. We shall do this by showing that $\varphi\alpha\varphi'^{-1}$ is the restriction to $\mathscr{A}(c+H)$ of the affinity $\mathscr{A}(f)$ (cf. exercise 1 of § 3.1).

In the notation of the Embedding Theorem (p. 32), if $S \in \mathscr{A}(c+H)$,

$$Sf = S' \quad \text{implies} \quad [S]f = [S'],$$

because f is a linear isomorphism. But f maps $c+H$ onto $c'+H'$. Also $S = [S] \cap (c+H)$ and $S' = [S'] \cap (c'+H')$. Thus

$$Sf = S' \quad \text{if, and only if,} \quad [S]f = [S'].$$

This is just another way of saying that

$$S\mathscr{A}(f) = S' \quad \text{if, and only if,} \quad S\varphi\alpha = S'\varphi',$$

i.e., $\varphi\alpha = \mathscr{A}(f)\varphi'$, as required.

Suppose, finally, that a projective isomorphism π induces an affinity α of A onto A′. Then $\varphi\alpha\varphi'^{-1}$ is of the form $\mathscr{A}(t_{-c}\, g\, t_{c'})$ where g is an isomorphism of H onto H'. We may extend g uniquely to an isomorphism f of V onto V' taking c onto c' (by the rule $(xc+h)f = xc' + hg$). The projectivity $\mathscr{P}(f)$ now clearly restricts to the affinity α and so we must have $\mathscr{P}(f) = \pi$.

Recapitulating, we have proved the following result.

PROPOSITION 3. *Let V, V' be vector spaces over fields F, F', respectively, and let H, H' be hyperplanes in $\mathscr{P}(V)$, $\mathscr{P}(V')$ determining affine geometries* A, A', *respectively. Then any projective isomorphism π of $\mathscr{P}(V)$ onto $\mathscr{P}(V')$ for which $H\pi = H'$ restricts to an affine isomorphism α of* A *onto* A'. *Conversely, any affine isomorphism α of* A *onto* A' *is the restriction of just one such projective isomorphism π.*

If $F = F'$ then π is a projectivity if, and only if, α is an affinity.

EXERCISE

1. Let P be a projective plane and A the affine plane determined by H as line at infinity. If π is a collineation of P leaving H invariant (i.e., $H\pi = H$) and α is the affinity of A obtained from π by restriction (according to Proposition 3), prove that α is a dilatation if, and only if, π is a central collineation with H as axis. Prove, further, that α is a translation if, and only if, π in addition has its center on H. (Cf. exercises 3 and 9 of § 3.2.)

3.5 Semi-linear Isomorphisms

Let V, V' be vector spaces over fields F, F', respectively and let α be an isomorphism of $\mathscr{A}(V)$ onto $\mathscr{A}(V')$. We also assume in the first place that $0_V\alpha = 0_{V'}$. For convenience we write $v\alpha = v'$ for all v in V.

When dim $V \geq 1$ we can choose a non-zero vector a in V and then, since α is an isomorphism taking the line $[a] = a \lrcorner 0_V$ onto the line $a' \lrcorner 0_{V'} = [a']$, α induces a one–one mapping of the points of $[a]$ onto the points of $[a']$. In other words, α defines a one–one mapping

$$\zeta : x \to x'$$

of F onto F' by the rule

$$(xa)' = x'a'$$

for all x in F.

Clearly, from this definition, $0_F' = 0_{F'}$ and $1_F' = 1_{F'}$, but as far as we can tell at this stage the mapping ζ depends on the choice of the non-zero vector a. In fact, when dim $V = 1$, an arbitrary one–one mapping of V onto V' induces an isomorphism.

When dim $V \geq 2$, however, we can deduce much more about both α and ζ.

Our first task is to show that ζ does not depend on the choice of the vector a in V. Suppose that a, b are linearly independent vectors in V and that

$$(xb)' = x''b'$$

for all x in F. It is clear that $0_F'' = 0_F' = 0_{F'}$ and so we may assume that $x \neq 0_F$. Now the line $xa \lrcorner xb$ is parallel to $a \lrcorner b$ and so, since α preserves parallelism, $x'a' \cup x''b'$ is parallel to $a' \cup b'$. But this implies immediately that $x'' = x'$ as required. If a and b are linearly dependent then we choose a third vector c not on $[a]$. By the above argument, the mappings of F onto F' determined by a, c and by b, c are the same, hence also those by a, b.

We may now say (when dim $V \geq 2$) that there is a *unique* mapping $\zeta: x \to x'$ of F onto F' such that

$$(xa)' = x'a' \tag{1}$$

for all x in F and all a in V.

If a and b are again linearly independent vectors in V then α carries the parallelogram $a, 0_V, b, a+b$ into the parallelogram $a', 0_{V'}, b', a'+b'$ and so

$$(a+b)' = a'+b'. \tag{2}$$

Let us now show that (2) is true for all a, b in V. Since it is clearly true if a or b is the zero vector, this only leaves the case where a, b are linearly dependent and non-zero. Choose a third vector c not on $[a]$. Then we may apply the rule to the pairs $(a+b, c)$, $(a, b+c)$, (b, c), so that

$$(a+b)' + c' = a' + (b+c)'$$
$$= a' + b' + c',$$

from which

$$(a+b)' = a' + b',$$

as required.

We may deduce two further properties of the mapping ζ from the rules (1) and (2).

First, choose any $a' \neq 0_{V'}$ and any x, y in F. Then

$$\begin{aligned}
(x+y)' \, a' &= ((x+y)a)' \\
&= (xa + ya)' \quad \text{(by the vector space axiom V.3,} \\
&\qquad\qquad\qquad\quad \text{applied to V)} \\
&= (xa)' + (ya)' \\
&= x'a' + y'a' \\
&= (x'+y')a' \quad \text{(by V.3, applied to V')}
\end{aligned}$$

and consequently

$$(x+y)' = x' + y' \tag{3}$$

for all x, y in F.

Finally, we invoke axiom V.4 which states that

$$(xy)a = x(ya)$$

for all x, y in F and all a in V. Thus

$$(xy)'a' = x'(ya)' = x'(y'a') = (x'y')a'$$

and so

$$(xy)' = x'y' \tag{4}$$

for all x, y in F.

DEFINITION. A one–one mapping ζ of a field F onto a field F' is called an *isomorphism* if

$$(x+y)\zeta = x\zeta + y\zeta,$$

$$(xy)\zeta = (x\zeta)(y\zeta)$$

for all x, y in F. If $F = F'$ then ζ is called an *automorphism* of F.

DEFINITION. Let V, V' be two vector spaces over fields F, F', respectively. If f is a one–one mapping of V onto V' and ζ is a one–one mapping of F onto F' where f and ζ are related by the rules

$$(a+b)f = af + bf,$$
$$(xa)f = (x\zeta)(af)$$

for all a, b in V and all x in F, then f is called a *semi-linear isomorphism* of V onto V' *with respect to* ζ.

We see from the proof of rules (3) and (4) given above, that the mapping ζ in the definition of a semi-linear isomorphism is *necessarily* an isomorphism of F onto F' (provided only that Vf contains a non-zero vector).

It is immediately obvious from the definition that the mapping $\mathscr{A}(f)$ induced by a semi-linear isomorphism f is an affine isomorphism. As a partial converse we have the following fundamental theorem.

THEOREM 5. *Let* $\mathsf{A} = \mathscr{A}(a+M)$, $\mathsf{A}' = \mathscr{A}(a'+M')$ *be affine geometries over fields* F, F', *respectively, and let* α *be an isomorphism of* A *onto* A'.
If dim $\mathsf{A} = 1$ *then there is a one–one mapping of* F *onto* F'.
If dim $\mathsf{A} \geq 2$ *then there is an isomorphism* ζ *of* F *onto* F' *and a semi-linear isomorphism* g *of* M *onto* M' *with respect to* ζ *such that*

$$\alpha = \mathscr{A}(t_{-a}\, g\, t_{a\alpha}).$$

The isomorphism ζ *and the semi-linear isomorphism* g *are both uniquely determined by* α.

PROOF. Very little remains to be proved. The mapping

$$\mathscr{A}(t_a)\, \alpha\, \mathscr{A}(t_{-a\alpha})$$

is an affine isomorphism of $\mathscr{A}(M)$ onto $\mathscr{A}(M')$ taking 0_M onto $0_{M'}$ and so, by the above argument, is induced by (i) a one–one mapping of F

onto F' when dim $A = 1$ and (ii) a semi-linear isomorphism g when dim $A \geq 2$.

The uniqueness of g is proved in exactly the same way as in the case when α is an affinity (cf. p. 43). But once g is known, ζ is uniquely determined by the argument given above.

Before we prove the corresponding theorem for projective geometries we remark that there is an analogue of Proposition 1 for semi-linear isomorphisms provided that dim $V \geq 2$. If f is a semi-linear isomorphism of V onto V' with respect to ζ and z' is any non-zero scalar in F' then the reader will verify that the mapping $g = z'f$ defined by the rule

$$vg = z'(vf)$$

for all v in V, is again a semi-linear isomorphism of V onto V' with respect to the *same* isomorphism ζ. The proof already given for Proposition 1 when dim $V \geq 2$ shows that the semi-linear isomorphisms f and g of V onto V' induce the same projective isomorphism $\mathscr{P}(f) = \mathscr{P}(g)$ if, and only if, $g = z'f$ for some non-zero scalar z' in F'.

THEOREM 6. *Let* $P = \mathscr{P}(V)$, $P' = \mathscr{P}(V')$ *be projective geometries over fields* F, F', *respectively and let* π *be an isomorphism of* P *onto* P'.

If pdim P $= 1$ *then there is a one–one mapping of* F *onto* F'.

If pdim P ≥ 2 *then there is an isomorphism* ζ *of* F *onto* F' *and a semi-linear isomorphism* f *of* V *onto* V' *with respect to* ζ *such that*

$$\pi = \mathscr{P}(f).$$

The isomorphism ζ *is uniquely determined by* π *and the semi-linear isomorphism* f *is determined by* π *to within an arbitrary (non-zero) scalar multiple in* F'.

PROOF. Choose a hyperplane H in $\mathscr{P}(V)$ and a vector c of V not in H. Let $H' = H\pi$ and c' be any non-zero vector in $[c]\pi$. The isomorphism π induces a mapping

$$\beta: S \to [S]\pi \cap (c' + H')$$

of $\mathscr{A}(c + H)$ onto $\mathscr{A}(c' + H')$ and this is an affine isomorphism (by part (3) of the Embedding Theorem, p. 32).

If pdim P $= 1$ then dim $\mathscr{A}(c + H) = 1$ and so the first part is immediate from Theorem 5.

If pdim P ≥ 2 then dim $\mathscr{A}(c + H) \geq 2$ and so there exists a field isomorphism ζ of F onto F' and a semi-linear isomorphism g of H onto H' with respect to ζ such that $\beta = \mathscr{A}(t_{-c} \, g \, t_{c'})$. We extend g uniquely to a semi-linear isomorphism f of V onto V' by the rule

$$(xc + h)f = (x\zeta)c' + hg.$$

Then the projective isomorphism $\mathcal{P}(f)$ induces β.   But π also induces β.   Hence $\mathcal{P}(f) = \pi$ by Proposition 3.

We recall that the fields we are mainly interested in are $\mathbf{F}_p$ (p prime), $\mathbf{Q}$, $\mathbf{R}$ and $\mathbf{C}$.   It is easy to show that $\mathbf{F}_p$ and $\mathbf{Q}$ each have only one automorphism, namely the identity.   Let ζ be an automorphism of $\mathbf{R}$. The positive elements of $\mathbf{R}$ are precisely the non-zero elements which are squares and so ζ preserves the natural order in $\mathbf{R}$.   Since ζ acts as the identity on the subfield $\mathbf{Q}$ of rational numbers, ζ preserves Dedekind sections and so is the identity on $\mathbf{R}$.   Thus the semi-linear isomorphisms are simply linear isomorphisms if the ground field F is $\mathbf{F}_p$, $\mathbf{Q}$ or $\mathbf{R}$. On the other hand $\mathbf{C}$ obviously has the automorphism $x + iy \to x - iy$ which is not the identity.   Surprisingly, $\mathbf{C}$ has a very large number of automorphisms—in fact $2^{2^{\aleph_0}}$ where $\aleph_0$ is the first transfinite cardinal.

EXERCISES

1. Prove that the identity mapping is the only automorphism of each of the fields $\mathbf{Q}$ and $\mathbf{F}_p$ for any prime p.   Show further that there are only two automorphisms of $\mathbf{C}$ that fix all the real numbers: the identity mapping and the conjugation mapping $x + iy \to x - iy$.

2. If ζ is an automorphism of the field F, prove that the mapping
$$g : (x_1, \ldots, x_n) \to (x_1\zeta, \ldots, x_n\zeta)$$
is a semi-linear isomorphism of F^n onto itself with respect to ζ.

3. If ζ is the conjugation mapping of exercise 1 and if g is defined as in exercise 2 in the case $F = \mathbf{C}$, show that a point P of $\mathcal{P}(\mathbf{C}^n)$ is fixed under $\mathcal{P}(g)$ if, and only if, P is a *real point*, i.e., P has a homogeneous vector in $\mathbf{R}^n$.

4. Let V be a real vector space and $V_{(\mathbf{C})}$ its complexification (cf. § 1.5, p. 14). If f is an isomorphism of V onto V and we define $g : V_{(\mathbf{C})} \to V_{(\mathbf{C})}$ to be the mapping
$$a + ib \to af - ibf$$
prove that g is a semi-linear isomorphism of $V_{(\mathbf{C})}$ onto $V_{(\mathbf{C})}$ with respect to $\zeta : x + iy \to x - iy$.

5. Let α be an isomorphism of $\mathscr{A}(V)$ onto $\mathscr{A}(V')$ and α' one of $\mathscr{A}(V')$ onto $\mathscr{A}(V'')$.   Suppose that $\dim V \geq 2$ and that α, α' determine semi-linear isomorphisms g, g' and field automorphisms ζ, ζ', respectively (according to Theorem 5).   Prove that $\alpha\alpha'$ determines the semi-linear isomorphism gg' with respect to $\zeta\zeta'$.

3.6   Groups of Automorphisms

In our discussion of isomorphisms (whether of vector spaces, affine geometries or projective geometries) we have not so far assumed that the image space coincides with the original. We shall now investigate what happens to our results when we do impose this restriction. The most notable difference is that it now *always* makes sense to multiply isomorphisms.

More precisely, let X be any one of the following objects: a vector space V, an affine geometry A, a projective geometry P or a field F. Then Aut X shall denote the set of all *automorphisms* of X, i.e., the set of all isomorphisms of X onto X. The identity mapping 1_X belongs to Aut X and if a, $b \in$ Aut X then ab and $a^{-1} \in$ Aut X. Moreover, mapping multiplication satisfies the associative law. Hence Aut X is actually a group with respect to multiplication. (It is a subgroup of the group of all permutations of X: cf. Chapter I, p. 4.)

The results of this chapter may now be reinterpreted as statements concerning the structure of the groups Aut V, Aut A and Aut P. That is the aim of this section.

We observe immediately that in Aut A there are two important subgroups: the subgroup Af A consisting of all affinities of A onto A, called the *affine group* on A, and the subgroup Tr A consisting of all translations in A. In Aut P there is the subgroup Pr P consisting of all collineations of P, called the *projective* (or *collineation*) *group* on P.

The group Aut V is usually called the *general linear group* on V. It is a subgroup of the group of all semi-linear isomorphisms of V onto V. We shall denote this latter group by $S(V)$.

We now need to introduce a new concept.

DEFINITION. A mapping θ of a group G into a group G' is a *homomorphism* if

$$(ab)\theta = (a\theta)(b\theta)$$

for every a, b in G. If θ is also one–one and onto G' then θ is an *isomorphism* of G onto G'.

In this definition we have written G and G' multiplicatively. It will be clear how the definitions must be phrased if other notations are used. Sometimes, in fact, one group is given with respect to addition and the other with respect to multiplication. We have a case of this at hand. If Tr V denotes the group of all translations t_a in V ($a \in V$), then the mapping

$$a \to t_a$$

is an isomorphism of the additive group V (see axiom V.1, p. 5) onto Tr V. This follows from the equation

$$t_{a+b} = t_a t_b.$$

If the homomorphism is not one–one we require some measure of the "shrinkage" when we go from G to the image $G\theta$.

DEFINITION. The *kernel* of the homomorphism θ of G into G' is the set of all elements a in G such that $a\theta = 1$ (the identity element of G').

PROPOSITION 4. *The kernel of θ is a subgroup of G.*

We leave the proof as an exercise. But we remark that not every subgroup need be a kernel. (See exercise 4, below.)

PROPOSITION 5. *Let V be a vector space of dimension ≥ 3 over a field F and let φ, ψ, θ be the homomorphisms*

$$F^* \xrightarrow{\varphi} \text{Aut } V \xrightarrow{\psi} \text{Aut } \mathscr{P}(V) \xrightarrow{\theta} \text{Aut} F$$

where $\varphi: z \rightarrow z1$, $\psi: f \rightarrow \mathscr{P}(f)$ and for any π in Aut $\mathscr{P}(V)$, $\pi\theta$ is the auto-morphism of F determined by π according to Theorem 6. Then

(i) *φ is one–one;*

(ii) *ψ has kernel $F^*\varphi$ and maps Aut$V$ onto the projective group Pr $\mathscr{P}(V)$;*

(iii) *θ has kernel Pr $\mathscr{P}(V)$ and maps Aut$\mathscr{P}(V)$ onto Aut$F$.*

PROOF. Clearly φ and ψ are homomorphisms. Moreover (i) is trivial and (ii) is immediate from Proposition 1.

To see that θ is a homomorphism, we choose two projective auto-morphisms and use Theorem 6 to represent them in the form $\mathscr{P}(f)$, $\mathscr{P}(f')$ where f, f' are semi-linear isomorphisms of V onto V with respect to ζ, ζ', respectively. Then $\mathscr{P}(f) \mathscr{P}(f') = \mathscr{P}(ff')$ and

$$(xa)(ff') = ((xa)f)f' = ((x\zeta)(af))f' = (x\zeta\zeta')(aff'),$$

so that ff' is semi-linear with respect to $\zeta\zeta'$. Hence θ is a homomorphism.

Finally, let ζ be an arbitrary element in Aut$F$. Choose an ordered basis $(a_1, \ldots, a_n)$ of V. Then the mapping

$$g: x_1a_1 + \cdots + x_na_n \rightarrow (x_1\zeta)a_1 + \cdots + (x_n\zeta)a_n$$

is a semi-linear isomorphism of V onto V with respect to ζ. Hence ζ is the image under θ of $\mathscr{P}(g)$.

The reader may be familiar with exact sequences of groups. Using this notion, Proposition 5 amounts to the statement that the sequence of groups

$$\{1\} \rightarrow F^* \xrightarrow{\varphi} \text{Aut } V \xrightarrow{\psi} \text{Aut } \mathscr{P}(V) \xrightarrow{\theta} \text{Aut } F \rightarrow \{1\}$$

is exact.

In order to describe efficiently the relations between the groups associated with an affine geometry we need the following concept.

DEFINITION. A group G is said to be the *split product* of the ordered pair of subgroups (H, K) if

(1) every element in G can be uniquely written in the form hk, with h in H, k in K and

(2) $hk \to h$ is a homomorphism (of G onto H with kernel K).

PROPOSITION 6. *Let V be a vector space of dimension ≥ 2 over a field F.*

(i) Aut $\mathscr{A}(V)$ *is the split product of (S, T) where S is isomorphic to $S(V)$ by means of the isomorphism $f \to \mathscr{A}(f)$ and $T = \mathrm{Tr}\, \mathscr{A}(V)$ is isomorphic to the additive group V, the isomorphism being $v \to \mathscr{A}(t_v)$.*

(ii) Af $\mathscr{A}(V)$ *is the split product of (L, T) where T is as before and L is isomorphic to Aut V, by means of the isomorphism $f \to \mathscr{A}(f)$, restricted to Aut V.*

The non-trivial part of this proposition is the proof of condition (1) for a split product in part (i): this follows from Theorem 5. The rest we leave as an exercise for the reader.

In view of the Embedding Theorem (p. 32), we expect some kind of relationship between Aut $\mathscr{P}(V)$ and Aut A. We may read this off from Proposition 3 above (p. 54).

THEOREM 7. *Let H be a hyperplane in $\mathscr{P}(V)$ and let A be the affine geometry in $\mathscr{P}(V)$ determined by H. Let $(\mathrm{Aut}\, \mathscr{P}(V))_H$ and $(\mathrm{Pr}\, \mathscr{P}(V))_H$ be the subgroups of Aut $\mathscr{P}(V)$ and Pr $\mathscr{P}(V)$ whose elements leave H invariant. Then the mapping $\rho : \pi \to \alpha$ defined in Proposition 3 is an isomorphism of $(\mathrm{Aut}\, \mathscr{P}(V))_H$ onto Aut A and ρ restricts to an isomorphism of $(\mathrm{Pr}\, \mathscr{P}(V))_H$ onto Af A.*

PROOF. The fact that $\rho : \pi \to \alpha$ preserves multiplication is immediate from the definitions. The rest of the theorem is immediate from Proposition 3. (Note again that in each of these statements the hardest part to prove is that the isomorphisms are *onto*.)

It is intuitively obvious that two isomorphic geometries will have isomorphic groups of automorphisms. We can make this quite explicit: If α is an isomorphism of an affine geometry A onto an affine geometry A′ then the mapping

$$\varphi \to \alpha^{-1}\varphi\alpha$$

is an isomorphism of Aut A onto Aut A′.

We may make use of the fact that A and $\mathscr{A}(H)$ are isomorphic to deduce that $(\mathrm{Pr}\, \mathscr{P}(V))_H$ is isomorphic to Af $\mathscr{A}(H)$. The reader should note, however, that the isomorphism here depends on the choice of coset $c + H$ and is no longer "natural".

EXERCISES

1. Given fields F, F' and a mapping $\varphi: F \to F'$ such that

$$(x+y)\varphi = x\varphi + y\varphi,$$

$$(xy)\varphi = (x\varphi)(y\varphi)$$

for all x, y in F; prove that φ is either an isomorphism of F onto $F\varphi$ or $x\varphi = 0_{F'}$, for all x in F.

2. If V is a vector space over F and dim $V = 1$, prove that Aut V is isomorphic to F^*.

3. If V is a vector space of dimension n over $\mathbf{F}_p$, prove that Aut V is finite and that its *order* (i.e., the number of its elements) is

$$(p^n - 1)(p^n - p)\ldots(p^n - p^{n-1}).$$

(Use exercise 3 of § 1.5 and Theorem 3 of Chapter I.)
Calculate the orders of Aut $\mathscr{P}(V)$ and Aut $\mathscr{A}(V)$.

4. Let G be the group of permutations of the set $\{1, 2, 3\}$. If π is the permutation defined by $1\pi = 2$, $2\pi = 1$, $3\pi = 3$, show that the subgroup $\{1, \pi\}$ of G cannot be the kernel of a homomorphism of G into another group. (If K is a kernel in a group G then $x^{-1}yx \in K$ for all y in K and all x in G.)

3.7 Central Collineations

It is perhaps surprising that the question of whether a given projective isomorphism is, or is not, a projectivity can already be decided by looking only at the effect it has on a single line.

LEMMA 2. *If π is a projective isomorphism whose restriction to a given line is known to be a projectivity, then π is itself a projectivity.*

PROOF. Let $\mathsf{P} = \mathscr{P}(V)$. If pdim $V = 1$ the result is obvious. If pdim $V \geq 2$ then $\pi = \mathscr{P}(f)$ where f is a semi-linear isomorphism with respect to ζ. Let g be the restriction of f to the given line L. Then $\mathscr{P}(g) = \mathscr{P}(h)$ for some *linear* isomorphism h of L. By the argument of Proposition 1, $g = zh$ for some non-zero scalar z. This implies that g is linear. But g is semi-linear with respect to ζ and so $\zeta = 1$ as required.

As a particular case we see that any projective automorphism π which has a line of invariant (or fixed) points (i.e., $P\pi = P$ for all points P on some line L) is a collineation.

DEFINITION. Let P be a projective geometry of dimension $n \geq 2$. Then any automorphism π of P which has a hyperplane H of invariant points is called a *central collineation*.

If H and K are distinct hyperplanes of invariant points for a col-

$x1_H$ and $y1_K$ respectively, for some scalars x, y. Since $H \cap K \neq 0$ we must have $x = y$. But then since P is the join of H and K, $\pi = \mathscr{P}(x1)$ is the identity collineation. Thus a central collineation π which is distinct from the identity collineation has a unique hyperplane of invariant points which we call *the hyperplane* of π.

If π has an invariant point A outside its hyperplane then every line through A contains at least two invariant points and so is left invariant by π. Thus $A, P, P\pi$ are collinear for any point P. We shall call a point A with this last property a *center* of π.

PROPOSITION 7. *A central collineation π which is distinct from the identity has one and only one center A (possibly on the hyperplane of π) and A is an invariant point for π.*

PROOF. Suppose that $\pi = \mathscr{P}(f)$ where f is an element of Aut V and that $P\pi = P$ for every point P of the hyperplane H of V. By multiplying f by a suitable non-zero scalar we may assume that the restriction of f to H is the identity mapping 1_H. Let c be a vector of V not in H; then any vector of V can be expressed uniquely in the form $xc + h$ where $x \in F$ and $h \in H$. By our assumption

$$(xc + h)f - (xc + h) = x(cf - c).$$

If $cf = c$ then of course f is the identity mapping on all of V. If $cf \neq c$ we may put $a = cf - c \neq 0$. The point $A = [a]$ is then a center as required.

If A and B are two distinct centers then $P\pi - P$ for all points P not on AB. But if C is an arbitrary point on AB and L is a line through C distinct from AB, then every point of L other than C is left invariant by π. Hence C itself is left invariant by π and so π is the identity collineation, contrary to hypothesis.

Finally, if A is the center of π, then every line through A is left invariant by π and so A is left invariant by π.

We could have defined a central collineation as one which has a center (see exercise 10, § 4.6, p. 83). The definition we chose is easier to handle although admittedly not so natural as far as terminology is concerned.

We recall from exercise 3 on page 44 that a *dilatation* of an affine geometry A is an affinity α of A onto A such that $S\alpha$ is parallel to S for every S in A.

PROPOSITION 8. *In the notation of Proposition 3, α is a dilatation of* A *if, and only if, π is a central collineation with hyperplane H. Further, α is a translation if, and only if, π has center on H. (Note that $\alpha = 1_A$ if, and only if, $\pi = 1_{\mathscr{P}(V)}$.)*

PROOF. (1) It follows directly from the definition that α is a dilatation if, and only if, S and $S\alpha$ have the same hyperplane at infinity for every S in A. But the construction of π from α given in the proof of Proposition 3 shows that this is equivalent to the condition that $M\pi = M$ for every subspace M of H.

(2) If α is a translation (other than the identity) then α has no invariant points. This implies that the invariant points of π, including the center, lie on H.

Conversely, suppose that $\pi \neq 1_{\mathscr{P}(V)}$ and has center on H. Then α has no invariant points (Proposition 7). The result now follows from exercise 9 of § 3.2. (The reader may prefer to verify that the mapping $\varphi \alpha \varphi^{-1}$ of $\mathscr{A}(c + H)$ onto itself is indeed a translation.) We may also argue that if P and Q are two distinct points of A then the points P, Q, $Q\alpha$, $P\alpha$ (in this order) are the vertices of a *parallelogram* (which is "flat" if PQ contains the center). But this is clearly a "geometrical" condition for α to be a translation.

A general collineation may be very different from a central collineation and in fact may not have *any* invariant points at all! Nevertheless we may prove the following striking result.

THEOREM 8. *Any collineation of a projective geometry* P *of dimension* $n \geq 2$ *can be expressed as a product of* n *(or fewer) central collineations of* P.

We shall need

LEMMA 3. *Given a subspace* M *of* V *and vectors* p, q *neither of which lies in* M, *then there exists an automorphism* f *of* V *such that* $pf = q$ *and that* f *is the identity mapping on a hyperplane* H *of* V *containing* M.

PROOF. Let $\{a_1, \ldots, a_r\}$ be a basis of M (which is empty if $M = 0$).

Suppose first that $M + [p] = M + [q]$. There exist linearly independent vectors $b_1, \ldots, b_s$ such that $\{a_1, \ldots, a_r, p, b_1, \ldots, b_s\}$ is a basis of V (Theorem 1 (3), of Chapter I, p. 10). Since $M + [p] = M + [q]$, $\{a_1, \ldots, a_r, q, b_1, \ldots, b_s\}$ is then also a basis. Hence there exists a unique automorphism f such that $a_i f = a_i$ for $i = 1, \ldots, r$, $b_j f = b_j$ for $j = 1, \ldots, s$ and $pf = q$ (Theorem 3 of Chapter I, p. 12).

If, on the other hand, $M + [p] \neq M + [q]$, then $M + [p] + [q]$ is a direct sum and we can complete its basis $\{a_1, \ldots, a_r, p, q\}$ to a basis of V by vectors $b_1, \ldots, b_{s-1}$. Here we define an automorphism f by $a_i f = a_i$ for $i = 1, \ldots, r$, $b_j f = b_j$ for $j = 1, \ldots, s-1$, $pf = q$ and $qf = p$. The hyperplane fixed elementwise by f is $[a_1, \ldots, a_r, b_1, \ldots, b_{s-1}, p + q]$.

PROOF OF THEOREM 8. For each π in Pr $\mathscr{P}(V)$ we consider the subspaces M on which π acts as the identity; and we define $m(\pi)$ to be the maximum of the dimensions pdim M for all such M. We shall prove that π can be written as a product of pdim $V - m(\pi)$ (or fewer) central collineations.

The proof is by induction on the integer $i(\pi) = \text{pdim}\, V - m(\pi)$. When $i(\pi) = 0$, π is the identity mapping which we regard conventionally as a product of no central collineations. When $i(\pi) = 1$, π is a central collineation. Suppose the result established for all π with $i(\pi) < k$ and consider a collineation with $i(\pi) = k$.

Let M be a subspace whose points are left invariant by π and with pdim $M = m(\pi)$. If P is any point not in M, then $P\pi$ also is not in M. The collineation π may be written $\pi = \mathscr{P}(g)$ where the restriction of g to M is 1_M. If $P = [p]$ and $q = pg$ then, by Lemma 3 above, there exists an automorphism f_1 such that f_1 restricts to the identity mapping on M, $q = pf_1$ and $\pi_1 = \mathscr{P}(f_1)$ is a central collineation. It follows that $f_1^{-1} g$ restricts to the identity mapping on $M + [q]$ and hence, by the induction hypothesis, $\mathscr{P}(f_1^{-1} g)$ is a product $\pi_2 \cdots \pi_k$ of central collineations. Thus $\mathscr{P}(g) = \mathscr{P}(f_1)\mathscr{P}(f_1^{-1} g) = \pi_1 \pi_2 \cdots \pi_k$, as required.

If π is a central collineation of $\mathscr{P}(V)$ with center A then the restriction of π to any hyperplane M of V is a projectivity of M onto $M\pi$. If A is not on M then this projectivity is called a *perspectivity* of M onto $M\pi$ with *center* A. It is clear that if P is any point of M then the corresponding point $P\pi$ is the intersection of $M\pi$ and the line AP. Conversely, it is quite easy to show that if two distinct hyperplanes M, M' in V are given, together with a point A not on M or M' and a hyperplane H through $M \cap M'$ but distinct from M and M', then there is a unique central collineation π with center A, hyperplane H and such that $M\pi = M'$.

In the foundations of classical projective geometry these perspectivities played the most important role. It is clear from Theorem 8 that a study of perspectivities leads to the general theory of projectivities, but that the projective isomorphisms induced by semi-linear isomorphisms, as distinct from linear isomorphisms, cannot be constructed in this way. As we pointed out earlier (page 58), the field **R** of real numbers has only the identity automorphism and so this distinction is vacuous in the truly classical case. The projective isomorphisms which arise over the complex numbers involving automorphisms of **C** other than the identity and the conjugacy automorphism are strictly "pathological" from the algebraic point of view.

CHAPTER IV

Linear Mappings

4.1 Elementary Properties of Linear Mappings

Our study of vector spaces in Chapter I involved only a single space at a time, except when we discussed isomorphisms. We must now look at relations between different spaces, possibly even of different dimensions. For this we need a generalization of the notion of isomorphism.

DEFINITION. If V and V' are vector spaces over the same field F, then a mapping f of V into V' is called a *linear mapping* (or *homomorphism*) if

    L.1. $(a+b)f = af + bf$,

    L.2. $(xa)f = x(af)$,

for all a, b in V and all x in F.

We sometimes wish to emphasise the linear nature of an isomorphism f of V onto V' and shall then refer to f as a *linear isomorphism*. We also recall that an isomorphism of V onto V itself is called a (*linear*) *automorphism* of V (cf. § 3.6, p. 59).

The most important objects associated with a linear mapping f of V into V' are the image Vf and a certain subset of V called the kernel of f.

DEFINITION. The *kernel* of f is the set of all vectors a in V such that $af = 0$, and will be denoted by $\mathrm{Ker}\, f$.

We leave the reader to prove that Vf is a subspace of V' and $\mathrm{Ker}\, f$ is a subspace of V. Further, f is one–one if, and only if, $\mathrm{Ker}\, f = \{0\}$. At the other extreme, f is the *zero mapping* 0 if, and only if, $\mathrm{Ker}\, f = V$.

THEOREM 1. *Let f be a linear mapping of the vector space V. Then for any subspace M of V such that*

$$V = M \oplus \mathrm{Ker}\, f$$

the restriction of f to M is an isomorphism of M onto Vf.

Moreover,

$$\dim Vf = \dim V - \dim (\mathrm{Ker}\, f).$$

PROOF. We recall from Chapter I, Theorem 1, Corollary 2 (p. 11) that there always exist subspaces M such that $V = M \oplus \text{Ker } f$. Any vector v in V is uniquely expressible in the form $v = a + b$ where $a \in M$ and $b \in \text{Ker } f$. By the definition of Ker f, $vf = af$ and so $Vf = Mf$. Further, $af = 0$ implies that $a \in M \cap \text{Ker } f = \{0\}$. Thus the restriction of f to M is an isomorphism of M onto $Mf = Vf$.

The last part now follows immediately from Theorem 2 of Chapter I (p. 11).

COROLLARY. *If f is a linear mapping of V into V', then any two of the following conditions imply the third:*

(1) *f is one–one;*
(2) *f is onto V';*
(3) *dim $V = $ dim V'.*

PROOF. (1) is equivalent to dim $V = $ dim Vf and (2) is equivalent to dim $Vf = $ dim V'.

Suppose again that $f: V \to V'$ is linear, $K = \text{Ker } f$ and $a, b \in V$. Then $af = bf$ if, and only if, $-a + b \in K$, i.e., if, and only if, a and b lie in the same coset of K (cf § 2.1). It is easy to see that if $u \in a + K$ and $v \in b + K$, then $u + v \in (a + b) + K$ and $xu \in xa + K$ for any scalar x. We may therefore define *unambiguously* an "addition" of cosets by setting $(a + K) + (b + K) = (a + b) + K$; and a "scalar multiplication" by setting $x(a + K) = xa + K$. If V/K denotes the set of all cosets of K, it is a simple matter to check that these definitions of addition and scalar multiplication make V/K into a vector space over F. Moreover, the mapping $p: a \to a + K$ of V onto V/K is linear and its kernel is precisely K.

Since f has the same effect on all the elements of a single coset of K, we may define a mapping p' of the space V/K by setting $(a + K)p' = af$. Then p' is linear and has image Vf. It is also one–one: if $(a + K)p' = (b + K)p'$ then $af = bf$ and hence $-a + b \in K$ so that $a + K = b + K$. We have now proved

PROPOSITION 1. *Every linear mapping f of V into V' can be factored in the form $f = pp'$, where p is the linear mapping $a \to a + \text{Ker } f$ of V onto $V/\text{Ker } f$ and p' is an isomorphism of $V/\text{Ker } f$ onto Vf.*

We observe that Theorem 1 can be expressed in a similar manner. If q is the mapping

$$q: a + b \to a$$

for a in M, b in Ker f, and q' is the restriction of f to M, then $f = qq'$. Since there are in general many "direct complements" M for Ker f in V, the factorization in this case is not unique. Despite this disadvantage we shall find more use in this book for Theorem 1 than for Proposition 1. The real significance of Proposition 1 is that the construction involved can be carried out in a much more general context than ours.

It is frequently necessary to consider more than one linear mapping at a time. It is then important to know that the set of all linear mappings (of a vector space into another) can itself be given the structure of a vector space.

Let $\mathscr{L}_F(V, V')$, or $\mathscr{L}(V, V')$ when the ground field F is understood, denote the set of all linear mappings of V into V'. If $f, g \in \mathscr{L}(V, V')$, we shall denote the mapping

$$a \to af + ag$$

by $f + g$; and for x in F, the mapping

$$a \to x(af)$$

by xf. We leave, as an exercise, the verification that $f + g$ and xf are themselves linear mappings (of V into V') and that, relative to these definitions of addition and scalar multiplication, $\mathscr{L}(V, V')$ is a vector space over F.

If $f \in \mathscr{L}(V, V')$ and $g \in \mathscr{L}(V', V'')$, then the mapping fg of V into V'' is again linear. (We recall that $a(fg) = (af)g$ for all a in V.) Further, if f and g are one–one (or isomorphisms) then fg is also one–one (or an isomorphism). The commutative law of multiplication $fg = gf$ need not hold even if both products fg, gf are defined (see exercises 2, 4, 5, below).

In the special case where $V = V'$, $\mathscr{L}(V, V)$ may be given more structure than just that of a vector space. For here the product fg of any two elements is always defined (and is an element of $\mathscr{L}(V, V)$). The following rules are easily checked:

A.1. $(fg)h = f(gh)$,
A.2. $f(g + h) = fg + fh$,
A.3. $(f + g)h = fh + gh$,
A.4. $x(fg) = (xf)g = f(xg)$,

for all f, g, h in $\mathscr{L}(V, V)$ and all x in F.

DEFINITION. Let A be a vector space over a field F together with a rule (called *multiplication*) which associates with any two elements a, b in A an element ab, also in A (and called the *product* of a by b). If the above axioms A.1 to A.4 hold for all f, g, h in A and all x in F, then A is called a *linear algebra over the field F*.

We recall that we have been assuming from Chapter II onwards that our vector spaces are finite dimensional. It is clear, however, that all the definitions given in this section (but not all the results!) are quite independent of this restriction.

Thus $\mathscr{L}(V, V)$, even in the infinite dimensional case, is an example of a linear algebra over F.   Another example is $F[X]$, the set of all polynomials in X over F (with the ordinary multiplication of polynomials).   A third important example arises if E is a field containing F: then E is a linear algebra over F, where the algebra multiplication is the given multiplication in E (cf. example 2 on p. 5).

EXERCISES

We drop the convention about finite dimensionality in exercises 1, 2, 4 *and* 9.

1. Prove that $f(X) \to f(X)^p$ is a one–one linear mapping of $\mathbf{F}_p[X]$ into itself.

2. If D denotes the operation of differentiation and J denotes the operation of "integration from 0 to X", show that D and J are linear mappings of $\mathbf{Q}[X]$ into $\mathbf{Q}[X]$.   Prove that D is onto $\mathbf{Q}[X]$ but is not one–one, and J is one–one but not onto $\mathbf{Q}[X]$.   (This shows that the corollary to Theorem 1 is false for infinite dimensional spaces.)   Find $JD - DJ$.

3. Let V be the vector space of all polynomials in X over a field F of degree less than n (including the zero polynomial).   If D denotes the differentiation operator applied to V, show that $\mathrm{Ker}\, D + VD$ is not the whole space V provided $n \geq 2$.

4. If D is the differentiation operator on $F[X]$ and m is the linear mapping $f(X) \to Xf(X)$, prove that $mD - Dm$ is the identity mapping.

5. If $(a_1, \ldots, a_m)$ is an ordered basis of U and if $b_1, \ldots, b_m$ are any vectors in V, show that there is one, and only one, linear mapping $f: U \to V$ such that $a_i f = b_i$, $i = 1, \ldots, m$.

    Let (e_1, e_2) be the standard basis of F^2 and let f and g be the linear mappings of F^2 into F^2 which take (e_1, e_2) to $(e_1, 0)$ and $(0, e_1)$, respectively.   Show that $fg = 0$ but that $gf = g$.

6. If U, V have ordered bases $(u_1, \ldots, u_m)$, $(v_1, \ldots, v_n)$, respectively, show that there is a linear mapping f_{ij} defined by

$$u_k f_{ij} = \delta_{ki} v_j \quad \text{for } i, k = 1, \ldots, m; j = 1, \ldots, n.$$

(Here δ_{ki} is the "Kronecker delta": its value is 1_F if $k = i$ and 0_F otherwise.)   Show that the mn mappings f_{ij} form a basis of $\mathscr{L}(U, V)$.

7. Let f be a linear mapping of V into V'.   If there exists a linear mapping g of V' into V such that either $fg = 1_V$ or $gf = 1_{V'}$, show that f is an isomorphism of V onto V' with inverse $f^{-1} = g$.   (Use exercise 1 of § 1.1 (p. 3) and Theorem 1, Corollary (p. 67).)

8. Vector spaces $V_0, V_1, \ldots, V_{n+1}$ are given together with linear mappings $f_i: V_i \to V_{i+1}$ for $i = 0, 1, \ldots, n$.   Suppose that $V_0 = V_{n+1} = 0$ and for $i = 0, \ldots, n-1$, $\mathrm{Ker}\, f_{i+1} = V_i f_i$.   (In these circumstances we say that the sequence

$$0 \longrightarrow V_1 \xrightarrow{f_1} V_2 \xrightarrow{f_2} \cdots \xrightarrow{f_{n-1}} V_n \longrightarrow 0$$

is *exact*.)   Prove that

$$\sum_{i=1}^{n} (-1)^i \dim V_i = 0.$$

9. Let f be a linear mapping of V into V satisfying $f^2 = f$. Show that if M is the image of f and K is the kernel of f, then $V = M \oplus K$. Hence show that f is the projection of V onto M with kernel K (cf. the paragraph following Proposition 1).

10. Let π be a non-identity central collineation of $\mathcal{P}(V)$ with hyperplane H and center A. Suppose that $\pi = \mathcal{P}(f)$, where the automorphism f is chosen so that f restricts to the identity mapping on H. Prove that $H = \mathrm{Ker}(f-1)$ and $A = V(f-1)$.

   (An automorphism f of V which restricts to the identity mapping on a hyperplane is called an *elementary transformation* of V. If $f \neq 1$, then the hyperplane of fixed vectors of f is necessarily $\mathrm{Ker}(f-1)$. Note also that, by Theorem 8 of Chapter III, p. 64, every automorphism of V can be written as a scalar times a product of elementary transformations. See also exercise 10 of § 4.6.)

11. If f is a non-identity elementary transformation with hyperplane H (for definitions see immediately above), show that the equation

$$(v + H)\bar{f} = vf + H$$

defines (unambiguously) an automorphism $\bar{f}$ of V/H. Prove that the central collineation $\mathcal{P}(f)$ has center on H if, and only if, $\bar{f}$ is the identity mapping on V/H.

   (An elementary transformation f for which $\bar{f}$ equals the identity, is called a *transvection* of V. Cf. the characterization of translations given in Proposition 8 of § 3.7, p. 63.)

4.2 Degenerate Affinities and Projectivities

If f is a linear mapping of V into V' then there is an induced mapping $\mathcal{A}(f)$ of $\mathcal{A}(V)$ into $\mathcal{A}(V')$ and an induced mapping $\mathcal{P}(f)$ of $\mathcal{P}(V)$ into $\mathcal{P}(V')$ (defined on pages 42, 44, respectively). When f is not one–one the mappings $\mathcal{A}(f)$ and $\mathcal{P}(f)$ also fail to be one–one. We shall stretch our usual terminology and refer to these mappings as *degenerate* affinities and projectivities; but an affinity or projectivity without this qualifying adjective is always understood to be non-degenerate.

If f has kernel $K \neq \{0\}$ then every subspace of K is mapped onto $\{0\}$ and so in particular every projective point in K (there may be only one!) is mapped onto $\{0\}$. In terms of the more classical notion of projective space (defined in § 2.5, p. 30), a point $[v]$ not in K has image point $[vf]$, but a point $[v]$ in K has no image point. In this case therefore f does not define a mapping of the underlying projective spaces. Nevertheless, the discussion of such mappings $\mathcal{P}(f)$ can be reduced to the case of a projectivity by using Theorem 1. For if M is any subspace of V such that $M \oplus K = V$, then the restriction of $\mathcal{P}(f)$ to the subgeometry $\mathcal{P}(M)$ is clearly a projectivity of $\mathcal{P}(M)$. Alternatively, we may use Proposition 1 to deduce that $\mathcal{P}(p')$ is a projectivity of $\mathcal{P}(V/K)$.

It is clear from the definitions of $\mathscr{A}(f)$ and $\mathscr{P}(f)$ that they preserve inclusion in the following sense:

$$S \subset T \quad \text{implies} \quad Sf \subset Tf.$$

$\mathscr{A}(f)$ also preserves parallelism, but may transform non-parallel cosets into parallel ones. Both $\mathscr{A}(f)$ and $\mathscr{P}(f)$ preserve joins because f itself does: more precisely, if $(a_i + M_i)_{i \in I}$ is a family of cosets, then

$$(\mathsf{J}(a_i + M_i : i \in I))f = \mathsf{J}(a_i f + M_i f : i \in I).$$

For, if each $a_i + M_i$ is actually a subspace (i.e., $a_i \in M_i$), then this relation is an immediate consequence of the definition "from below" of the join (sum) of subspaces given in Chapter I (p. 7); while the general result follows from this case together with the identity

$$\mathsf{J}(a_i + M_i : i \in I) = a_{i_0} + \mathsf{J}([a_i - a_{i_0}], M_i : i \in I)$$

(cf. Lemma 3, p. 18). Neither $\mathscr{A}(f)$ nor $\mathscr{P}(f)$ however, need preserve intersections (cf. exercise 1 below).

EXERCISES

1. Let f be the linear mapping $(x, y, z) \rightarrow (0, x+y, 0)$ of F^3 into itself and let $M = [(1, 0, 0)]$, $N = [(0, 1, 0)]$. Show that $(M \cap N)f \neq Mf \cap Nf$.

2. If f and g are linear mappings such that $\mathscr{P}(f) = \mathscr{P}(g)$, prove that $g = zf$ for some non-zero scalar z.

3. If g is a linear mapping of V into itself such that $\mathscr{P}(g)^2 = \mathscr{P}(g)$, prove that there exists a linear mapping f of V into itself such that $\mathscr{P}(f) = \mathscr{P}(g)$ and $f^2 = f$. Describe all such "projections" $\mathscr{P}(g)$ when pdim $V = 3$. (Cf. exercise 9, § 4.1.)

4. Let f be a linear mapping of V into V and, for any scalar x, let $E_x = \text{Ker}(f - x1)$. If $E_x \neq \{0\}$ we call x an *eigenvalue* of f and E_x the corresponding *eigenspace*. The non-zero elements of E_x are called *eigenvectors*. Let $\mathscr{P}(f)$ be a (possibly degenerate) collineation. Show that a point $[v]$ is an invariant point of $\mathscr{P}(f)$ if, and only if, v is an eigenvector of f. Prove further that two distinct eigenspaces of f have no points in common.

4.3 Matrices

If V and V' are vector spaces over F and f is a linear mapping of V into V', then we may find "coordinates" of f as follows.

Choose ordered bases $(v_1, \ldots, v_m)$, $(v_1', \ldots, v_n')$ of V, V', respectively. Then the equations

$$v_i f = \sum_{j=1}^{n} a_{ij} v_j', \quad i = 1, \ldots, m,$$

define the set of coefficients a_{ij} in F uniquely.

DEFINITION. A rectangular array of elements of F with m rows and n columns is called an $m \times n$ *matrix*. For fixed F, m, n the set of all such matrices is denoted by $F^{m \times n}$.

The matrix

$$\begin{pmatrix} a_{11} & a_{12} & \cdot & \cdot & \cdot & a_{1n} \\ a_{21} & a_{22} & \cdot & \cdot & \cdot & a_{2n} \\ \cdot & \cdot & \cdot & \cdot & \cdot & \cdot \\ a_{m1} & a_{m2} & \cdot & \cdot & \cdot & a_{mn} \end{pmatrix}$$

is often written (a_{ij}). The element a_{ij} is called the (i, j)-*term* of the matrix.

The i-th row of the matrix (a_{ij}) appearing in the above equations is the coordinate row of $v_i f$ with respect to the ordered basis $(v_1', \ldots, v_n')$. We say that (a_{ij}) is the *matrix of f with respect to the ordered bases* (v_i), (v_j') and write

$$(a_{ij}) = (f; (v_i), (v_j')),$$

or, when there is no risk of confusion, $(a_{ij}) = (f)$.

Once the ordered bases $(v_1, \ldots, v_m)$, $(v_1', \ldots, v_n')$ of V, V' are given, the mapping $f \to (f) = (a_{ij})$ is a one–one mapping of $\mathscr{L}(V, V')$ onto $F^{m \times n}$ (cf. § 4.1 exercise 6). We now define three operations on matrices to conform with this one–one mapping.

DEFINITION. If (a_{ij}), $(b_{ij}) \in F^{m \times n}$ then

(1) $(a_{ij}) + (b_{ij}) = (a_{ij} + b_{ij})$;

(2) $x(a_{ij}) = (xa_{ij})$, for x in F.

If $(a_{ij}) \in F^{m \times n}$, $(b_{jk}) \in F^{n \times p}$ then

(3) $(a_{ij})(b_{jk}) = (c_{ik}) \in F^{m \times p}$, where

$$c_{ik} = \sum_{j=1}^{n} a_{ij} b_{jk}.$$

In view of parts (1), (2) of the definition, the one–one mapping of $\mathscr{L}(V, V')$ onto $F^{m \times n}$ preserves the operations of addition and multiplication by scalars and hence $F^{m \times n}$ is a vector space over F. As such it is isomorphic to $\mathscr{L}(V, V')$. The matrix corresponding to the zero mapping is called the *zero matrix* and will be denoted by 0.

Observe that $F^{1 \times n}$ is naturally isomorphic to F^n. We shall identify these two spaces.

The rule of multiplication for matrices gives

PROPOSITION 2. *If $f \in \mathscr{L}(U, V)$, $g \in \mathscr{L}(V, W)$ and (u_i), (v_j), (w_k) are ordered bases of U, V, W, respectively, then*

$$(f; (u_i), (v_j))(g; (v_j), (w_k)) = (fg; (u_i), (w_k));$$

or, more loosely,

$$(f)(g) = (fg).$$

PROOF. If $u_i f = \sum_{j=1}^{n} a_{ij} v_j$ and $v_j g = \sum_{k=1}^{p} b_{jk} w_k$, then

$$u_i(fg) = \sum_{k=1}^{p} \sum_{j=1}^{n} a_{ij} b_{jk} w_k = \sum_{k=1}^{p} c_{ik} w_k.$$

Certain properties of matrices now follow immediately from the corresponding properties of linear mappings. For example, the associative law

$$(AB)C = A(BC)$$

of matrix multiplication follows from the associative law which is true for mappings in general. (Of course the matrices A, B, C must be "conformable" for multiplication, i.e., the number of columns of A, B must equal the number of rows of B, C, respectively.)

We denote by I_n, or by I if no confusion is possible, the $n \times n$ matrix (δ_{ij}) (where δ_{ij} is the Kronecker delta (cf. exercise 6 § 4.1), i.e., I has 1's down the "diagonal" $i = j$ and 0's elsewhere. If V has ordered basis $(a_1, \ldots, a_n)$ then I is the matrix $(1_V; (a_i), (a_j))$ of the identity mapping 1_V on V. Therefore it is clear that if $A \in F^{m \times n}$ and $B \in F^{n \times p}$ then

$$A I = A \quad \text{and} \quad I B = B.$$

I is usually called the *identity matrix*.

A more striking example of the use of Proposition 2 is the following. If $A = (f; (u_i), (v_j))$, where f is an isomorphism of U onto V, then we may define

$$A^{-1} = (f^{-1}; (v_j), (u_i)).$$

By Proposition 2,

$$A A^{-1} = I = A^{-1} A$$

and so we call A^{-1} an *inverse* of A.

The matrix A^{-1} seems to depend on the particular isomorphism of which it is the matrix (not to mention the spaces U, V and the bases (u_i), (v_j)), but in fact such an inverse, if it exists, is necessarily unique. For if

$$L A = I \quad \text{or} \quad A R = I$$

we have

$$L A A^{-1} = I A^{-1} \quad \text{or} \quad A^{-1} A R = A^{-1} I$$

from which

$$L = A^{-1} \quad \text{or} \quad R = A^{-1}.$$

DEFINITION. A square matrix A is *invertible* (or *non-singular*) if there exists a (necessarily square) matrix A^{-1} satisfying

$$A A^{-1} = A^{-1} A = I.$$

Then A^{-1} is called the *inverse* of A.

PROPOSITION 3. *A square matrix A is invertible if, and only if, A is the matrix of an isomorphism.*

The "if" part was established above and the "only if" part we leave as an exercise.

EXERCISES

1. If A, B are square matrices such that $AB=I$, prove that A, B are both invertible and that each is the inverse of the other. (Cf. exercise 7, § 4.1.)

2. Let E_{ij} denote the $m \times n$ matrix with all terms zero except the (i, j)-term, which is 1 $(i=1, \ldots, m;\ j=1, \ldots, n)$. Prove that these E_{ij}'s form a basis of $F^{m \times n}$. We call this the *standard basis of $F^{m \times n}$*.

   (Cf. exercise 6 of § 4.1. Observe also that if $m=1$ and we identify $F^{1 \times n}$ with F^n, then the standard basis in the above sense coincides with the standard basis $\{e_1, \ldots, e_n\}$ of F^n as defined earlier: exercise 9 of § 1.4.)

3. Let V be the vector space of all polynomials in $F[X]$ of degree less than n (including the zero polynomial). If D is the differentiation operator applied to V, find the matrix
   $$(D; (1, X, \ldots, X^{n-1}), (1, X, \ldots, X^{n-1})).$$

4. Let A be a square matrix such that $A^k=0$ for some positive integer k. (Such a matrix A is called *nilpotent*.) Prove that $I-A$ is invertible with inverse
   $$I+A+A^2+\cdots+A^{k-1}.$$

5. Let A be the matrix

   $$\begin{pmatrix} 4 & 3 & 0 \\ 0 & 1 & 0 \\ 3 & 0 & -2 \end{pmatrix}$$

   with coefficients in $\mathbf{F}_p$ and let V be the subspace of $\mathbf{F}_p^{3 \times 3}$ spanned by all the powers of A (including $A^0 = I$). Find $\dim_{\mathbf{F}_p} V$ when p equals 2, 3 and 5.

6. If f is the mapping of $\mathbf{C}$ into $\mathbf{R}^{2 \times 2}$ given by

   $$f: a+ib \rightarrow \begin{pmatrix} a & b \\ -b & a \end{pmatrix},$$

   prove that f is a one–one linear mapping of real vector spaces and that

   $$(xy)f = (xf)(yf)$$

   for all x, y in $\mathbf{C}$. (Thus, in particular, f is an isomorphism of the field $\mathbf{C}$.)

7. Let D be the subset of $\mathbf{C}^{2 \times 2}$ consisting of all matrices of the form

$$\begin{pmatrix} x & y \\ -\bar{y} & \bar{x} \end{pmatrix}$$

where $x, y \in \mathbf{C}$ and the bar denotes complex conjugation. Prove that D is a linear algebra over $\mathbf{R}$ but not over $\mathbf{C}$ (cf. p. 68). Show further that D is a division ring but is not a field (cf. p. 39). (D is called the *real quaternion algebra*.)

8. If $A = (a_{ij}) \in F^{n \times n}$, let At denote the sum of the diagonal elements of A: i.e.,

$$At = a_{11} + \cdots + a_{nn}.$$

(We call At the *trace* of A and t the *trace mapping*.) Prove that t is a linear mapping of $F^{n \times n}$ onto F and verify that

$$(AB)t = (BA)t$$

for all A, B in $F^{n \times n}$.

9. If f is a linear mapping of $F^{n \times n}$ into F such that

$$(AB)f = (BA)f$$

for all A, B in $F^{n \times n}$, prove that $f = xt$ for some x in F and where t denotes the trace as in the last exercise. (Consider the effect of f on the standard basis of $F^{n \times n}$ defined in exercise 2 above.)

10. (1) Let H be a hyperplane in the n-dimensional projective geometry $\mathscr{P}(V)$ and π a collineation leaving H invariant. Choose an ordered basis $(v_0, \ldots, v_n)$ of V so that $H = [v_1, \ldots, v_n]$. Show that there exists an automorphism f of V such that $\pi = \mathscr{P}(f)$ and

$$(f; (v_i), (v_j)) = \begin{pmatrix} 1 & a_1 & a_2 & \cdots & a_n \\ 0 & & & & \\ 0 & & A & & \\ \vdots & & & & \\ 0 & & & & \end{pmatrix}$$

where $a_1, \ldots, a_n \in F$ and A is an invertible $n \times n$ matrix.

　　(2) Show that the set of all such matrices (f) is a group G isomorphic to $(\mathrm{Pr}\ \mathscr{P}(V))_H$ (in the notation of Theorem 7, Chapter III, p. 61).

　　(3) Let g denote the restriction of f to H and let $a = \sum_{i=1}^{n} a_i v_i$. Show that the mapping

$$g\, t_a \to (f)$$

is an isomorphism of $\mathrm{Af}.\mathscr{A}(H)$ onto G.

11. In the notation of exercise 10 (3) above, find the subgroup of G which corresponds to the group of dilatations of $\mathscr{A}(H)$. Hence prove Proposition 8 of Chapter III, p. 63.

4.4　The Rank of a Linear Mapping

　　We have seen that matrices arise naturally in describing linear mappings. They also arise when there is a change of basis.

Let $(a_1, \ldots, a_m)$ and $(b_1, \ldots, b_m)$ be two ordered bases of V. Then we may express each a_i as a unique linear combination of the b_j's:

$$a_i = \sum_{j=1}^{m} p_{ij} b_j, \quad i = 1, \ldots, m,$$

or, in terms of the notation introduced above,

$$P = (p_{ij}) = (1_V; (a_i), (b_j)).$$

This shows at once that P is invertible with inverse

$$P^{-1} = (1_V; (b_j), (a_i)).$$

If the vector v has coordinate rows $x = (x_1, \ldots, x_m)$, $y = (y_1, \ldots, y_m)$ with respect to the ordered bases (a_i), (b_j), respectively, then we easily prove

$$y = xP.$$

PROPOSITION 4. *If $f \in \mathscr{L}(V, V')$ has matrices A, B with respect to two pairs of bases of V, V', then there exist invertible matrices P, Q such that $B = P A Q$.*

PROOF. Let $A = (f; (v_i), (v_j'))$ and $B = (f; (a_r), (a_s'))$. If we put $P = (1_V; (a_r), (v_i))$ and $Q = (1_{V'}; (v_j'), (a_s'))$, then $B = P A Q$.

Given a linear mapping f of V into V' we naturally wish to choose bases of V and V' yielding as simple a matrix for f as possible.

THEOREM 2. *If $f \in \mathscr{L}(V, V')$, then we can choose ordered bases (a_i), (a_j') of V, V' such that*

$$(f; (a_i), (a_j')) = \begin{pmatrix} I_r & 0 \\ 0 & 0 \end{pmatrix}$$

where $r = \dim Vf$.

PROOF. If we choose an ordered basis $(a_1, \ldots, a_m)$ of V such that $(a_{r+1}, \ldots, a_m)$ is an ordered basis of Ker f, then, by Theorem 1, $(a_1 f, \ldots, a_r f)$ is an ordered basis of Vf. Put $a_j' = a_j f$, $j = 1, \ldots, r$, and extend $(a_1', \ldots, a_r')$ to an ordered basis $(a_1', \ldots, a_n')$ of V'. The matrix $(f; (a_i), (a_j'))$ has then the required form.

DEFINITION. *If $f \in \mathscr{L}(V, V')$, we call $\dim Vf$ the rank of f.*

Observe that the rank of f is at most $\dim V$; and that it equals $\dim V$ if, and only if, f is one–one.

In order to give Theorem 2 a purely matrix formulation we introduce a further definition.

DEFINITION. The maximum number of linearly independent rows of a matrix A is called the *row rank* of A. The maximum number of linearly independent columns of A is the *column rank of A*.

COROLLARY. *If $A \in F^{m \times n}$, then there exist invertible matrices P, Q such that*

$$P A Q = \begin{pmatrix} I_r & 0 \\ 0 & 0 \end{pmatrix}$$

where r is the row rank of A.

PROOF. The result follows from Proposition 4 and Theorem 2 since the rank of a linear mapping equals the row rank of any matrix representing it.

DEFINITION. Two matrices A, B in $F^{m \times n}$ are called *equivalent* if there exist invertible matrices P, Q such that $B = P A Q$.

We check that this is indeed an equivalence relation. The above corollary shows that A, B *are equivalent if, and only if, they have the same row rank.*

In order to clarify the connexion between the rows and columns of a matrix we introduce the transpose of a matrix.

DEFINITION. The *transpose A^t* of the $m \times n$ matrix A is the $n \times m$ matrix whose (i, j)-term is the (j, i)-term A. In other words, A^t is obtained from A by interchanging the rows and columns. The matrix A is *symmetric* if $A^t = A$.

It follows at once from the definition that $(A + B)^t = A^t + B^t$, and $(xA)^t = xA^t$. We also have the less obvious

PROPOSITION 5. $(A B)^t = B^t A^t$.
The proof is left as an exercise.

From the symmetry of the matrix $P A Q = \begin{pmatrix} I_r & 0 \\ 0 & 0 \end{pmatrix}$

it is easy to prove

THEOREM 3. *The row and column ranks of a matrix are equal.*
For we have

$$\begin{aligned}
\text{row rank of } A &= \text{row rank of } P A Q \\
&= \text{row rank of } Q^t A^t P^t \\
&= \text{row rank of } A^t \\
&= \text{column rank of } A.
\end{aligned}$$

DEFINITION. The *rank* of a matrix A is its row, or column, rank. It will be denoted by $r(A)$.

We give an alternative proof of Theorem 3, independent of the theory of linear mappings.

Let r, c be the row and column ranks of A. Without loss of generality the first r rows of A are linearly independent and so all the rows of A are linearly dependent on the first r. In other words there exist scalars x_{ik} such that

$$a_{ij} = \sum_{k=1}^{r} x_{ik} a_{kj} \quad \text{for} \quad i = 1, \ldots, m \text{ and } j = 1, \ldots, n.$$

This may be written

$$A = \begin{pmatrix} x_{11} & \cdots & x_{1r} \\ x_{21} & \cdots & x_{2r} \\ \cdot & \cdot & \cdot \\ x_{m1} & \cdots & x_{mr} \end{pmatrix} \begin{pmatrix} a_{11} & \cdots & a_{1n} \\ a_{21} & \cdots & a_{2n} \\ \cdot & \cdot & \cdot \\ a_{r1} & \cdots & a_{rn} \end{pmatrix},$$

from which it is clear that each column of A is linearly dependent on the columns of (x_{ik}); but there are only r of these, and so $c \leq r$. The reverse inequality can be proved by a similar argument and so $c = r$.

PROPOSITION 6. *If $A, B \in F^{n \times n}$ then*

(1) $r(A + B) \leq r(A) + r(B)$;
(2) $r(AB) = r(BA) = r(A)$ *if B is invertible*;
(3) $r(AB) \leq \min(r(A), r(B))$; *and*
(4) $r(AB) \geq r(A) + r(B) - n$.

PROOF. These results all follow quickly if A, B are interpreted as matrices of linear mappings f, g of V into V. We shall only prove (4).

Let h be the restriction of the mapping g to the subspace Vf. Then Ker $h \subset$ Ker g, and so, by Theorem 1,

$$\dim Vf - \dim Vfh \leq \dim V - \dim Vg.$$

But $Vfh = Vfg$, by the definition of h, so that

$$r(f) - r(fg) \leq n - r(g),$$

which is a reformulation of the statement (4).

If we are given a linear mapping f of V into itself and (a_i) an ordered basis of V, then we are naturally interested in the matrix

$$A = (f; (a_i), (a_j)).$$

We shall refer to A as the *matrix of f with respect to the ordered basis* (a_i). The problem of finding (a_i) so that A has as simple a form as possible is considerably more difficult than the equivalence problem we solved in Theorem 2. We shall consider this question in detail in the last chapter, but we note here that if $B = (f; (b_j), (b_k))$ and $P = (1_V; (a_i), (b_j))$, then $B = P^{-1} A P$.

DEFINITION. Two matrices A, B in $F^{n \times n}$ are called *similar* if there exists an invertible matrix P such that $B = P^{-1} A P$.

The reader may check that similarity is an equivalence relation on $F^{n \times n}$. The problem to be studied in the last chapter may therefore be stated in matrix terms as follows: given a square matrix A, to find the simplest matrix similar to A.

EXERCISES

1. If $A \in F^{m \times n}$ and if $x A y^t = 0$ for all x in $F^{1 \times m}$ and y in $F^{1 \times n}$, show that $A = 0$.
   If $S \in F^{n \times n}$ and if $x S x^t = 0$ for all x in $F^{1 \times n}$ show that $S = -S^t$. (Such a matrix S is called *skew-symmetric*.)

2. If F is a field which is not of characteristic 2 and if $A \in F^{n \times n}$, show that the linear mapping $x \to xA$ has the property $(xA)(xA)^t = x \, x^t$ for all x in F^n if, and only if, $A A^t = I$. What can be deduced about A if F is a field of characteristic 2? (A square matrix A for which $A A^t = I$ (and so necessarily $A^t A = I$ by exercise 1 of § 4.3) is called an *orthogonal* matrix.)

3. If $S \in F^{n \times n}$ show that the matrices $I + S$, $I - S$ commute. If F is not of characteristic 2 and if $I - S$ is invertible, show that

$$A = (I+S)(I-S)^{-1} = (I-S)^{-1}(I+S)$$

is orthogonal (exercise 2) if, and only if, S is skew-symmetric (exercise 1).

4. If $A \in F^{m \times n}$, $B \in F^{n \times p}$ show that

$$\text{rank } AB \geq \text{rank } A + \text{rank } B - n.$$

5. Show that a matrix C in $F^{m \times p}$ has rank r if, and only if, C can be expressed as a product $C = A B$ where $A \in F^{m \times r}$, $B \in F^{r \times p}$ and A, B both have rank r. What does this become if $r = 1$?

4.5 Linear Equations

The system of linear equations

$$\sum_{j=1}^{n} a_{ij} X_j = b_i, \quad i = 1, \ldots, m$$

may be written in the convenient notation

$$A X = b$$

where X stands for the $n \times 1$ matrix $(X_1, \ldots, X_n)^t$ and b for the $m \times 1$ matrix $(b_1, \ldots, b_m)^t$.

In this notation it is clear, for example, that if A is invertible then the equations have the unique solution $A^{-1}b$. (As a practical method of finding the inverse A^{-1} of a given invertible matrix A, one may solve the equations $AX = b$ with "indeterminates" b by the method of successive elimination explained in § 3.3 and thus exhibit A^{-1} in the form $X = A^{-1}b$.)

We wish to give an alternative proof of Theorem 4 of Chapter III (p. 50) using Theorem 1 of the present chapter. In order to do this we transpose the above matrix equation (with $b = 0$) and find

$$X^t A^t = 0^t.$$

We may now interpret A^t as the matrix of a linear mapping f of F^n into F^m (referred to the standard bases). The kernel of f is then precisely the set of all solutions $(x_1, \ldots, x_n)$ and by Theorem 1 it has dimension $(n - r)$ where r is the rank of f. But r is the row rank of A^t (cf. the proof of the corollary to Theorem 2) and so r is the column rank of A. This is the same as the row rank of A (Theorem 3) and so Theorem 4 of Chapter III is proved again.

We shall meet another proof in the next section.

4.6 Dual Spaces

Since F may be regarded as a one-dimensional vector space over F, the set $\mathscr{L}(V, F)$ is a special case of $\mathscr{L}(V, V')$ and is thus a vector space over F.

DEFINITION. We write $V^* = \mathscr{L}(V, F)$ and call this the *dual space* of V. The elements of V^* are called *linear forms* on V.

The following important, but simple, result is a special case of exercise 6, § 4.1.

THEOREM 4. *Given an ordered basis $(a_1, \ldots, a_n)$ of V, there is a unique ordered basis $(e_1, \ldots, e_n)$ of V^* such that*

$$a_i e_j = \delta_{ij}.$$

DEFINITION. $(e_1, \ldots, e_n)$ is called the *dual basis* of $(a_1, \ldots, a_n)$. We shall adopt the notation $e_i = a^i$.

Observe that if v has coordinate row $(x_1, \ldots, x_n)$ with respect to $(a_1, \ldots, a_n)$ and f is a linear form with coordinate row $(y_1, \ldots, y_n)$ with respect to the dual basis $(a^1, \ldots, a^n)$, then $vf = x_1 y_1 + \cdots + x_n y_n$.

COROLLARY 1. *The vector spaces V, V^* have the same dimension.*

COROLLARY 2. *If v is a non-zero vector in V then there is an element f in V^* such that $vf \neq 0$.*

For any f in V^*, v in V, vf is a scalar in F. If f is fixed as v runs through V, the values vf define the function $f: v \to vf$. If, however, v is fixed and f runs through V^*, the values vf define the function

$$v\iota : f \to vf.$$

It is easily shown that $v\iota$ is a linear form on V^*; in other words, ι is a mapping of V into V^{**}.

THEOREM 5. *The mapping ι is an isomorphism of V onto V^{**}.*

PROOF. Since

$$(a+b)\iota: \ f \to (a+b)f = af+bf,$$
$$(xa)\iota: \ f \to (xa)f = x(af),$$

the mapping ι is linear. It is one–one by Corollary 2, Theorem 4. Finally, $\dim V = \dim V^* = \dim V^{**}$, by Corollary 1, Theorem 4, and so, by Corollary 1, Theorem 1, $V\iota = V^{**}$.

Since the isomorphism ι does not depend on any choice of basis of V but is uniquely defined as soon as V is known, we shall frequently *identify v with $v\iota$* and thus identify V with V^{**}.

Although V and V^* are isomorphic (since they have the same finite dimension), there is in general no such unique isomorphism of V onto V^*. (The next chapter is largely concerned with a study of the isomorphisms of V onto V^*.) There does, however, exist a natural one–one mapping of the subspaces of V onto those of V^*.

DEFINITION. If M is a subspace of V then M° shall denote the set of all f in V^* such that $af = 0$ for all a in M. We call M° the *annihilator* of M and $\circ$ the *annihilator mapping*.

It follows at once that M° is a subspace of V^*. Further, in terms of the identification introduced above, if P is a subspace of V^* then P° is the set of all a in V such that $af = 0$ for all f in P.

THEOREM 6. *The mapping $M \to M^\circ$ is a one–one mapping of the subspaces of V onto the subspaces of V^* and has the following properties:*
(1) $\dim M^\circ = \dim V - \dim M$;
(2) $M^{\circ\circ} = M$;
(3) $M \subset N$ *if, and only if,* $M^\circ \supset N^\circ$;
(4) $(M+N)^\circ = M^\circ \cap N^\circ$;
(5) $(M \cap N)^\circ = M^\circ + N^\circ$.

PROOF. Let $(a_1, \ldots, a_r)$ be an ordered basis of M and extend it to an ordered basis $(a_1, \ldots, a_n)$ of V. Let $(a^1, \ldots, a^n)$ be the dual basis. Now $f \in M^\circ$ if, and only if, $a_1 f = \cdots = a_r f = 0$. If $f = x_1 a^1 + \cdots + x_n a^n$, this means precisely that $x_1 = \cdots = x_r = 0$. Hence M° is the subspace $[a^{r+1}, \ldots, a^n]$ of V^*. This gives (1).

By definition, $af = 0$ for all f in $M°$ and all a in M, so that $M \subset M°°$. But by (1) applied to $M°$ (using the identification ι), we have

$$\dim M°° = \dim V^* - \dim M° = \dim M,$$

whence $M°° = M$.

If $M° = N°$, then $M°° = N°°$, i.e., $M = N$, by (2). Further, if P is a subspace of V^*, then $P = (P°)°$. We have thus shown that the mapping $M \to M°$ is one–one and onto the set of all subspaces of V^*. It remains to check the properties (3), (4), (5).

(3) Clearly $M \subset N$ implies $M° \supset N°$; and $M° \supset N°$ implies $M°° \subset N°°$, i.e., $M \subset N$, by (2).

(4) We have $(a + b)f = 0$ for all a in M, b in N, if, and only if, $af = 0$, $bf = 0$ for all a in M, b in N, and this is exactly (4).

(5) Applying (4) to $M° + N°$, we see that

$$(M° + N°)° = M°° \cap N°° = M \cap N,$$

by (2), whence $M° + N° = (M \cap N)°$.

The first and most immediate application of Theorem 6 is to give yet another proof of Theorem 4 of Chapter III.

We have the equations

$$\sum_{j=1}^{n} a_{ij}X_j = 0, \quad i = 1, \ldots, m.$$

If V is the vector space F^n and the coefficient space is M, then the solution space is isomorphic to $M°$. Property (1) of Theorem 6 now gives the required result.

EXERCISES

The definition of the dual space V^* and of the mapping ι apply equally well when V is infinite dimensional.

We drop our convention of finite dimensionality in exercises 1, 3, 5 and 6.

1. Show that if $x \in F$ then the mapping $p(X) \to p(x)$ is a linear form on $F[X]$.

2. Show that the mapping $f_i : (x_1, \ldots, x_n) \to x_i$ is a linear form on F^n and that $(f_1, \ldots, f_n)$ is the dual basis of the standard basis of F^n.

3. If S is any subset of V and $S°$ is the set of all f in V^* such that $sf = 0$ for all s in S, prove that $S°$ is a subspace of V^* and that $S° = M°$ where M is the subspace of V spanned by S.

4. If $(M_i)_{i \in I}$ is a family of subspaces of a vector space V, prove that

     (i) $(+(M_i : i \in I))° = \bigcap(M_i° : i \in I)$,

     (ii) $(\bigcap(M_i : i \in I))° = +(M_i° : i \in I)$.

5. If $V = F[X]$ show that the equations

$$x^j f_i = \delta_{ij} \quad (i, j = 0, 1, 2, \ldots)$$

define elements $f_0, f_1, f_2, \ldots$ of V^* and that the subspace they span is strictly smaller than V^*.

6. If $V = F[X]$ show that the mapping ι of V into V^{**} is linear and one–one. Assuming that $\{f_0, f_1, f_2, \ldots\}$ can be extended to a basis of V^*, prove that ι is not onto V^{**}.

7. If $f \in \mathscr{L}(U, V)$ and $h \in \mathscr{L}(V, F)$ then $fh \in \mathscr{L}(U, F)$. Hence the equation $hf^t = fh$ defines a mapping f^t of V^* into U^*. Show that $f^t \in \mathscr{L}(V^*, U^*)$ (f^t is called the *transpose* of f). Show that if U^{**}, V^{**} are identified with U, V by the isomorphism of Theorem 5, then $(f^t)^t$ is identified with f. Hence prove that the mapping $f \to f^t$ is a one–one mapping of $\mathscr{L}(U \ V)$, onto $\mathscr{L}(V^*, U^*)$. Show also that $(f+g)^t = f^t + g^t$, $(xf)^t = xf^t$ and $(fg)^t = g^t f^t$.

8. If A is the matrix of f in $\mathscr{L}(U, V)$, with respect to the ordered bases (u_i), (v_j), show that A^t is the matrix of f^t (see exercise 7) with respect to the dual bases of (v_j), (u_i), respectively.

9. Let $V = F^{n \times n}$ and $t: V \to F$ denote the trace mapping defined in exercise 8 of § 4.3. Show that the mapping

$$f_B: A \to (AB)t,$$

where $A, B \in V$, is a linear form on V. Show further that

$$B \to f_B$$

is an isomorphism of V onto V^*. (Use the standard basis of $F^{n \times n}$.)

10. Let A be a subspace of the vector space V and let f be a semi-linear isomorphism (with respect to ζ) of V onto itself for which $Af = A$. Show that the equation

$$(v + A)\bar{f} = vf + A$$

defines *unambiguously* a semi-linear isomorphism $\bar{f}$ of V/A onto itself with respect to the same automorphism ζ.

Suppose that dim $V \geq 3$ and that $\pi = \mathscr{P}(f)$ is a projective isomorphism of $\mathscr{P}(V)$ onto itself for which a *center* exists, i.e., there is a point $A = [a]$ such that A, P, $P\pi$ are collinear for every point P in $\mathscr{P}(V)$. Show that the induced mapping $\mathscr{P}(\bar{f})$ is the identity mapping on $\mathscr{P}(V/A)$. Hence show that

   (i) f is linear, i.e., π is a collineation;

   (ii) $vf = zv + ya$ (v in V), where z is a fixed scalar and $v \to y$ is a linear form on V;

   (iii) either π is the identity collineation or π has a hyperplane H of fixed points where H is the kernel of $v \to y$ (so that, in either case, π is a central collineation).

4.7   Dualities

The annihilator mapping $M \to M^\circ$ gives us a one–one mapping of $\mathscr{P}(V)$ onto $\mathscr{P}(V^*)$. It is a *natural* link between the two geometries although, of course, it is not an isomorphism. Indeed, it is almost the opposite: the mapping and its inverse both *reverse* the inclusion relation (Theorem 6, (3)). This has important consequences for the geometry $\mathscr{P}(V)$. The most immediate is the "principle of duality" (Theorem 7, below), which essentially "doubles" the theorems at our disposal without our having to do any extra work.

A proposition $\mathfrak{P}$ in n-dimensional projective geometry over the field F is a statement involving only the subspaces of the geometry and inclusion relations between them. It is usually phrased in terms of intersections, joins and dimensions. We define the *dual proposition* $\mathfrak{P}^*$ to be the statement obtained from $\mathfrak{P}$ by changing $\subset$ to $\supset$ throughout and hence replacing intersection, join and dimension r by join, intersection and dimension $n-1-r$, respectively.

Similarly, if $\mathfrak{C}$ is a configuration in an n-dimensional projective geometry over F then the *dual configuration* $\mathfrak{C}^*$ is obtained from $\mathfrak{C}$ by reversing all inclusion signs and hence replacing intersection, join and dimension r by join, intersection and dimension $n-1-r$, respectively.

For example, when $n=2$ and $\mathfrak{C}$ is a *complete quadrangle*, viz., the configuration consisting of four points, no three of which are collinear, and the six lines joining them in pairs, then the dual configuration is a *complete quadrilateral* and consists of four lines, no three of which are concurrent, and their six points of intersection.

We note that $\mathfrak{P} = \mathfrak{P}^{**}$ and $\mathfrak{C} = \mathfrak{C}^{**}$ for all propositions $\mathfrak{P}$ and all configurations $\mathfrak{C}$.

THEOREM 7. (THE PRINCIPLE OF DUALITY.) *If $\mathfrak{P}$ is a proposition which is true in all n-dimensional projective geometries over a given field F, then $\mathfrak{P}^*$ is also true in all n-dimensional projective geometries over F.*

PROOF. Let $\mathscr{P}(V)$ be a projective geometry over F of dimension n and suppose that the conditions of the proposition $\mathfrak{P}^*$ are satisfied in $\mathscr{P}(V)$. Then, by Theorem 6, the conditions of proposition $\mathfrak{P}^{**} = \mathfrak{P}$ are satisfied in $\mathscr{P}(V^*)$. Hence, by our hypothesis, the conclusions of $\mathfrak{P}$ are true in $\mathscr{P}(V^*)$. Again by Theorem 6, the conclusions of $\mathfrak{P}^*$ are true in $\mathscr{P}(V)$.

The importance of the annihilator mapping suggests the following concept.

DEFINITION. If δ is a one–one mapping of $\mathscr{P}(V)$ onto $\mathscr{P}(W)$ such that

$$M \subset N \quad \text{if, and only if,} \quad M\delta \supset N\delta$$

for all M, N in $\mathscr{P}(V)$, then δ is called an *anti-isomorphism* of $\mathscr{P}(V)$ onto $\mathscr{P}(W)$.

It is obvious that the mapping product $\delta \circ : \mathscr{P}(V) \to \mathscr{P}(W) \to \mathscr{P}(W^*)$ is then an isomorphism of $\mathscr{P}(V)$ onto $\mathscr{P}(W^*)$. If dim $V \geq 3$ we may apply Theorem 6 of the last chapter (p. 57) and deduce that $\delta \circ = \mathscr{P}(f)$, where f is a semi-linear isomorphism of V onto W^*. We may write this equally well as $\delta = \mathscr{P}(f)\circ$, where now $\circ$ denotes the annihilator mapping from $\mathscr{P}(W^*)$ onto $\mathscr{P}(W)$ (using Theorem 5 to identify $\mathscr{P}(W^{**})$ with $\mathscr{P}(W)$).

DEFINITION. If f is a *linear* isomorphism of V onto W^* then $\mathscr{P}(f)\circ$ is called a *duality* of $\mathscr{P}(V)$ onto $\mathscr{P}(W)$. A duality of $\mathscr{P}(V)$ onto itself is called a *correlation* of $\mathscr{P}(V)$.

Thus, for example, a correlation of a 3-dimensional projective geometry maps points to planes and planes to points and maps lines to lines.

Let f be a linear mapping (not necessarily an isomorphism) of V into V^*. If (a_i) is an ordered basis of V and (a^i) denotes, as usual, the dual basis of V^*, then we call

$$A = (f; (a_i), (a^j))$$

the *matrix of* f *with respect to* (a_i).

We wish to calculate how A changes when we switch to a new basis (b_i) of V. Let

$$R = (1_V; (a_i), (b_j)) \quad \text{and} \quad S = (1_{V^*}; (a^i), (b^j)).$$

Then, if

$$B = (f; (b_i), (b^j)),$$

we have

$$B = R^{-1} A S.$$

So it remains to find the relation between R and S.

If $x = (x_i)$ is the coordinate row of v in V with respect to (a_i) and $p = (p_i)$ is the coordinate row of h in V^* with respect to (a^i), then

$$vh = \sum_i x_i p_i = xp^t.$$

If y, q are the coordinate rows of v, h referred to (b_i), (b^i), respectively, then

$$y = xR \quad \text{and} \quad q = pS.$$

But

$$xp^t = vh = yq^t$$

and so

$$xp^t = xRS^t p^t$$

for all x, p. We deduce that $RS^t = I$, whence $R^{-1} = S^t$. Thus we have proved

PROPOSITION 7. *If $f \in \mathscr{L}(V, V^*)$ and A, B are the matrices of f with respect to two ordered bases of V, then*

$$B = S^t A S$$

for some invertible matrix S.

DEFINITION. Two matrices A, B in $F^{n \times n}$ are called *congruent* if there exists an invertible matrix S such that $B = S^t A S$.

It is simple to check that congruence is an equivalence relation on $F^{n \times n}$. We shall be much concerned in the next chapter with the problem of finding the simplest matrix to represent a given linear mapping of V into V^*. Proposition 7 shows that this is essentially the same as searching for the simplest matrix congruent to a given one.

EXERCISES

1. Show that the dual of Desargues' Theorem in the projective plane is the converse. What is the dual of Desargues' Theorem in a three dimensional geometry?

2. Find the dual of exercise 1, § 2.7 (p. 41).

3. The configuration $\mathfrak{C}$ consisting of all the points on a line is called a *range of points*. Describe the plane dual and the three-dimensional dual of $\mathfrak{C}$ (called a *pencil of lines* and a *pencil of planes*, respectively).
   Find the three-dimensional dual of the configuration consisting of all points on a projective plane.

4.8 Dual Geometries

In the last section we viewed the annihilator mapping as a connexion between the geometries $\mathscr{P}(V)$ and $\mathscr{P}(V^*)$. It can also be regarded in another way. The definition of generalized projective geometry given in Chapter II (p. 35) shows that the set of all subspaces of V together with the mapping $\circ$ is a projective geometry. We shall denote this geometry by $\mathscr{P}^*(V)$ and call it the *dual geometry* of $\mathscr{P}(V)$.

The first thing to note is that $\mathscr{P}(V)$ and $\mathscr{P}^*(V)$ are identical *as sets*; but their geometrical structures are very different. Here is one of the cases mentioned on p. 35 where care must be taken to say in exactly which context the inclusion, intersection, join and dimension are to be understood. For example, if M, N are subspaces of V then their intersection and join as members of the geometry $\mathscr{P}(V)$ are $M \cap N$, $M + N$, respectively, but their intersection and join as members of $\mathscr{P}^*(V)$ are $M + N$, $M \cap N$, respectively. If M has projective dimension r as member of $\mathscr{P}(V)$ then M has projective dimension $n-1-r$ as member of $\mathscr{P}^*(V)$ (where pdim $V = n$).

It is clear that a correlation of $\mathscr{P}(V)$ may be interpreted as a projectivity of $\mathscr{P}(V)$ onto $\mathscr{P}^*(V)$; and more generally, a duality of $\mathscr{P}(V)$ onto $\mathscr{P}(W)$ may be regarded as a projectivity of $\mathscr{P}(V)$ onto $\mathscr{P}^*(W)$.

In the terminology of exercise 3 of § 4.7, a *range of points* is a projective space of dimension one, i.e., the set of all points on a projective line. If pdim $V = 2$ then a range of points in $\mathscr{P}^*(V)$ is the set of all lines through a fixed point of $\mathscr{P}(V)$, i.e., a *pencil of lines* in $\mathscr{P}(V)$. When pdim $V = 3$ a range of points in $\mathscr{P}^*(V)$ is the set of all planes through a fixed line in $\mathscr{P}(V)$, i.e., a *pencil of planes* in $\mathscr{P}(V)$. This makes precise what is meant by saying that a pencil of lines or a pencil of planes is a one-dimensional projective space. One can apply to them all the familiar theory of one-dimensional projective geometry.

The relationships between the subspaces of $\mathscr{P}(V)$ and $\mathscr{P}^*(V)$ can often be most easily discussed in terms of the "dual coordinates". Let $(A_0, \ldots, A_n, U)$ be a frame of reference for $\mathscr{P}(V)$ and $(a_0, \ldots, a_n)$ any basis determining it (in the sense that $[a_i] = A_i$ for all i and $[\sum a_i] = U$: cf. p. 46). Then the dual basis $(a^0, \ldots, a^n)$ determines a frame of reference for $\mathscr{P}(V^*)$ and this, using the link $\circ$, yields a frame of reference $(A_0', \ldots, A_n', U')$ for $\mathscr{P}^*(V)$. Note that, by the last part of Lemma 1 of Chapter III (p. 47), the frame $(A_0', \ldots, A_n', U')$ is uniquely determined by $(A_0, \ldots, A_n, U)$.

DEFINITION. We call $(A_0', \ldots, A_n', U')$ the *dual frame of reference* of $(A_0, \ldots, A_n, U)$.

It is possible to give a geometric construction of $(A_0', \ldots, A_n', U')$, as opposed to the algebraic one used above. For each $i = 0, \ldots, n$, we know that $(A_i')^\circ = [a^i]$ and from this we see immediately that $A_i' = \oplus (A_j : j \neq i)$, i.e., A_i' is the hyperplane face of the simplex opposite A_i. The construction of U', on the other hand, is somewhat more involved (see exercise 1 below).

Let us examine how the dual coordinates of a point P in $\mathscr{P}^*(V)$ may be found. Since P° is a point in $\mathscr{P}(V^*)$ (by the definition of $\mathscr{P}^*(V)$), we know that P is a hyperplane in $\mathscr{P}(V)$. By Proposition 2 of Chapter III (p. 51), P is determined by a single homogeneous linear equation $p_0 X_0 + \cdots + p_n X_n = 0$ with respect to the frame $(A_0, \ldots, A_n, U)$. Therefore $p_0 a^0 + \cdots + p_n a^n$ is a homogeneous vector for P° and consequently P, as a point in $\mathscr{P}^*(V)$, has projective coordinate row $(p_0, \ldots, p_n)$ with respect to $(A_0', \ldots, A_n', U')$. This fact is frequently expressed by saying that the hyperplane P has dual coordinates $(p_0, \ldots, p_n)$ with respect to $(A_0, \ldots, A_n, U)$.

Finally, here is a simple example to illustrate the use of dual coordinates. Suppose that pdim $V = 2$ and that A, L are a point and line in $\mathscr{P}(V)$, A not lying on L. We should like to prove that the mapping $P \to AP$ is a *projectivity* of the range of points on L onto the pencil with vertex A. Take as triangle of reference ABC where B, C are distinct points of L (and some suitable point as unit point). Then the point P of L has coordinates $(0, p_1, p_2)$, say, and the corresponding line AP has equation $p_2 x_1 - p_1 x_2 = 0$. The dual coordinates of AP are $(0, p_2, -p_1)$. The *linear* nature of the mapping

$$(p_1, p_2) \to (p_2, -p_1)$$

establishes the fact that $P \to AP$ is a projectivity.

EXERCISES

1. Let $(A_0, \ldots, A_n, U)$ be a frame of reference for the projective geometry $\mathscr{P}(V)$ and $(A_0', \ldots, A_n', U')$ the dual frame of reference. We have seen that A_i' is the hyperplane face of the simplex $(A_0, \ldots, A_n)$ opposite A_i.

    (1) If $n = 1$ show that U' (which is a point) is the harmonic conjugate of U with respect to A_0, A_1.

    (2) Assume that $n \geq 2$ and let U_n be the point $A_n U \cap A_n'$. Show that $(A_0, \ldots, A_{n-1}, U_n)$ is a frame of reference for the subgeometry $\mathscr{P}(A_n')$ and that the dual frame of reference is

    $$(A_0' \cap A_n', \ldots, A_{n-1}' \cap A_n', U' \cap A_n').$$

    (3) Give an inductive construction for U' in terms of the frame of reference $(A_0, \ldots, A_n, U)$.

2. A pencil of lines L with vertex A in a projective plane M is given; a point B not in M is also given. Show that the mapping $L \to L \cup B$ is a projectivity of the given pencil of lines onto the pencil of planes with axis AB.

3. Two coplanar pencils with distinct vertices A, B, and a projectivity of one pencil onto the other, are given. Show that the points of intersection of pairs of corresponding lines are collinear if, and only if, AB is a self-corresponding line. (Cf. exercise 7 of § 3.2.)

4. In a projective plane two distinct lines L, M are given. If $X \to X'$ is a projectivity of L onto M, show that all the intersections $PQ' \cap P'Q$ (for $P \neq Q'$, $P' \neq Q$, $P \neq Q$) lie on a fixed line (called the cross-axis of the projectivity). [Hint: If R, $R' \neq L \cap M$ consider the pencils RX', $R'X$ as X varies on L; use exercise 3 above to show that the points $RX' \cap R'X$ lie on a line N_R. Finally show that N_R is independent of R.]

5. Deduce Pappus' Theorem from exercise 4.

CHAPTER V

Bilinear Forms

5.1 Elementary Properties of Bilinear Forms

DEFINITION. Let V be a vector space over a field F. A *bilinear form* σ on V is a mapping of the ordered pairs of vectors of V into F such that

B.1. $\sigma(xa + yb, c) = x\sigma(a, c) + y\sigma(b, c)$,
B.2. $\sigma(a, xb + yc) = x\sigma(a, b) + y\sigma(a, c)$,

for all vectors a, b, c in V and all scalars x, y in F.

We express condition B.1 by saying that σ is linear in the first variable and condition B.2 by saying that σ is linear in the second variable.

In this chapter we shall be concerned with the structure of V as a vector space with a given bilinear form σ. Under these circumstances we shall speak of V, or more exactly of (V, σ), as a *bilinear space*.

DEFINITION. An *isomorphism* of the bilinear space (V, σ) onto the bilinear space (V', σ') is a linear isomorphism f of V onto V' which satisfies

$$\sigma(a, b) = \sigma'(af, bf)$$

for all a, b in V. If such an isomorphism f exists we shall say that (V, σ) and (V', σ') are *isomorphic*.

If (V, σ) and (V, σ') are isomorphic then σ and σ' are said to be *congruent*.

If f is a linear mapping of V into V we denote by σ^f the bilinear form obtained from σ by the rule

$$\sigma^f(a, b) = \sigma(af, bf)$$

for all a, b in V. Therefore σ and σ' are congruent bilinear forms on V if, and only if, there exists an automorphism f of V such that $\sigma = (\sigma')^f$, or $\sigma' = \sigma^{f^{-1}}$. The reader will easily verify that congruence is an equivalence relation on the set $\mathscr{B}(V)$ of all bilinear forms on V.

We obtain a natural vector space structure on $\mathscr{B}(V)$ if we define $\sigma + \tau$ and $x\sigma$ as follows:

$$(\sigma + \tau)(a, b) = \sigma(a, b) + \tau(a, b),$$
$$(x\sigma)(a, b) = x\,\sigma(a, b)$$

for any σ, τ in $\mathscr{B}(V)$ and any x in F.

Suppose that an ordered basis $(a_1, \ldots, a_n)$ of V is given and that the vectors a, b have coordinate rows $x = (x_1, \ldots, x_n)$, $y = (y_1, \ldots, y_n)$, respectively, with respect to this ordered basis. Then, by the bilinearity of σ,

$$\sigma(a, b) = \sum s_{ij} x_i y_j,$$

where $s_{ij} = \sigma(a_i, a_j)$. If S is the $n \times n$ matrix (s_{ij}), we may rewrite this equation in matrix form as

$$\sigma(a, b) = x\,S\,y^t.$$

We call S the matrix of σ with respect to $(a_1, \ldots, a_n)$ and shall use the notation $(\sigma; (a_i)) = S$. The polynomial $\sum s_{ij} X_i Y_j$ is called the *bilinear polynomial* of σ with respect to $(a_1, \ldots, a_n)$. The following result is now immediate from the definitions.

PROPOSITION 1. *Let $(a_1, \ldots, a_n)$ be an ordered basis of V over F.*
(1) *The mapping $\sigma \to (\sigma; (a_i))$ is a linear isomorphism of $\mathscr{B}(V)$ onto $F^{n \times n}$.*
(2) *The mapping $\sigma \to \sum \sigma(a_i, a_j) X_i Y_j$ is a linear isomorphism of $\mathscr{B}(V)$ onto the vector space of all bilinear polynomials in $X_1, \ldots, X_n; Y_1, \ldots, Y_n$.*

Suppose that σ, σ' are congruent bilinear forms on V and $(a_1, \ldots, a_n)$ is an ordered basis of V. Let $S = (s_{ij}) = (\sigma; (a_i))$, $T = (t_{ij}) = (\sigma'; (a_i))$ and $a_i f = \sum_k p_{ik} a_k$ $(i = 1, \ldots, n)$. By the definition of σ' and the bilinearity of σ,

$$t_{ij} = \Big(\sum_k \sum_l p_{ik} p_{jl} s_{kl}\Big),$$

i.e., $T = P\,S\,P^t$, where $P = (p_{ij})$. Thus S, T are congruent matrices in the sense of § 4.7 (p. 86). Conversely, if σ, τ are bilinear forms such that $(\sigma; (a_i))$, $(\tau; (a_i))$ are congruent matrices, then σ, τ are congruent. This follows by reversing the above argument. We have therefore established

PROPOSITION 2. *Let $(a_1, \ldots, a_n)$ be an ordered basis of V. The bilinear forms σ, τ are congruent if, and only if, the matrices $(\sigma; (a_i))$, $(\tau; (a_i))$ are congruent.*

COROLLARY. *The matrices of a fixed bilinear form σ with respect to all ordered bases of V form a complete congruence class.*

PROOF. Given two ordered bases $(a_1, \ldots, a_n)$, $(b_1, \ldots, b_n)$ there is a unique f in $\mathrm{Aut}\, V$ such that

$$b_i = a_i f \quad (i = 1, \ldots, n).$$

Now $(\sigma^f; (a_i)) = (\sigma; (b_i))$, by the definition of σ^f, while $(\sigma^f; (a_i))$, $(\sigma; (a_i))$ are congruent by Proposition 2. Thus the matrices of σ with respect to all ordered bases lie in a single congruence class. To prove that we obtain the whole class, we take any matrix T congruent to $(\sigma; (a_i))$. By Proposition 1, $T = (\tau; (a_i))$ for some bilinear form τ and by Proposition 2, $\tau = \sigma^f$ for some automorphism f of V. Hence $T = (\sigma; (b_i))$ where $b_i = a_i f$ $(i = 1, \ldots, n)$.

The following definition is now unambiguous.

DEFINITION. Let σ be a bilinear form on a vector space V. If $(a_1, \ldots, a_n)$ is any ordered basis of V then the *rank* of σ is the rank of $(\sigma; (a_i))$. If $\mathrm{rank}\,\sigma < n$ we say that σ is *degenerate* and if $\mathrm{rank}\,\sigma = n$ we say that σ is *non-degenerate*.

We shall now describe a fundamental connexion between the theory of bilinear forms and the theory of linear mappings. The most satisfactory treatment of bilinear forms is based on a full exploitation of this relation.

Let σ belong to $\mathscr{B}(V)$. Since σ is linear in the second variable, the mapping

$$b \to \sigma(a, b)$$

is a linear form on V. If we denote this form by $a\underset{\sim}{g}$, then, since σ is linear in the first variable, $\underset{\sim}{g}$ is a linear mapping of V into the dual space V^*. Conversely, if f is any element of $\mathscr{L}(V, V^*)$, we may define a bilinear form σ on V by the rule

$$\sigma(a, b) = b(af).$$

If we interchange the roles of the two variables we can define an element $\tilde{\sigma}$ of $\mathscr{L}(V, V^*)$ by the rule

$$\sigma(a, b) = a(b\tilde{\sigma}),$$

i.e.,

$$b\tilde{\sigma} : a \to \sigma(a, b).$$

PROPOSITION 3. *The mappings $\sigma \to \underset{\sim}{g}$, $\sigma \to \tilde{\sigma}$ defined by the equations*

$$\sigma(a, b) = b(a\underset{\sim}{g})$$
$$\sigma(a, b) = a(b\tilde{\sigma})$$

(for all a, b in V), are linear isomorphisms of $\mathscr{B}(V)$ onto $\mathscr{L}(V, V^*)$.

PROOF. The mapping $f \to \sigma$, constructed above, is the inverse of $\sigma \to g$ and hence $\sigma \to g$ is one–one and onto $\mathscr{L}(V, V^*)$. The linearity follows immediately from the definitions of $\sigma + \tau$ and $x\sigma$.

If $(a_1, \ldots, a_n)$ is again an ordered basis of V and $s_{ij} = \sigma(a_i, a_j)$ then

$$a_i g : a_j \to s_{ij}$$

and so $a_i g$ is the linear form $\sum_j s_{ij} a^j$. Thus the matrix

$$(g; (a_i), (a^j))$$

is simply (s_{ij}). (Observe that the corollary to Proposition 2 is now an immediate consequence of Proposition 3 and Proposition 7 of Chapter IV (p. 86)).

In exactly the same way we see that

$$(\tilde{\sigma}; (a_i), (a^j)) = (s_{ij})^t.$$

In view of the equality of row and column ranks for a matrix (Theorem 3 of Chapter IV, p. 77), the ranks of g and $\tilde{\sigma}$ are both equal to the rank of σ.

EXERCISES

In the first three exercises the vector spaces in question are not assumed to be finite dimensional. The definitions for an infinite dimensional vector space of a bilinear form σ and of the mappings g, $\tilde{\sigma}$ are exactly as given above.

1. Consider the following properties of a bilinear form σ:
   (i) $\sigma(a, b) = 0$ for all b implies $a = 0$;
   (ii) $\sigma(a, b) = 0$ for all a implies $b = 0$.
   Prove that (i) is equivalent to g being one–one and (ii) is equivalent to $\tilde{\sigma}$ being one–one.
   If V is finite dimensional show that (ii) is also equivalent to g being onto V^*.
   (Thus in the case of a finite dimensional vector space V, each of these conditions is equivalent to the non-degeneracy of σ. For general V it is usual to take (i) and (ii) together as defining the non-degeneracy of σ.)

2. Let (s_{ij}) be a given family of elements in a field F, indexed by the set of all ordered pairs of non-negative integers. Prove that there is one and only one bilinear form on $F[X]$ such that

   $$\sigma(X^i, X^j) = s_{ij}$$

   for all $i, j \geq 0$ (where $X^0 = 1$).

3. If σ is the bilinear form on $F[X]$ such that

   $$\sigma(X^i, X^j) = \delta_{i+1, j}$$

   for all $i, j \geq 0$ (Kronecker delta), prove that g is one–one but that $\tilde{\sigma}$ is not one–one. (Cf. exercise 1 above.)

4. If $\sigma \in \mathscr{B}(V)$ and $f \in \mathscr{L}(V, V)$, prove that

$$(\underset{\sim}{\sigma}') = f\sigma f^t$$

where f^t is the transpose of f, as defined in exercise 7 of § 4.6 (p. 83).

5.2   Orthogonality

DEFINITION.   If σ is a bilinear form on a vector space V and if $\sigma(a, b) = 0$ then we say that a, b (in that order) are *orthogonal*, or that a is orthogonal to b (with respect to σ). Two subsets A, B of V are *orthogonal* if a, b are orthogonal for every a in A, b in B.

If M is any subspace of V then we denote by $M^{\perp(\sigma)}$ the set of all vectors b in V such that $\sigma(M, b) = 0$ (i.e., for which $\sigma(a, b) = 0$ for all a in M); and by $M^{\top(\sigma)}$ the set of all vectors a in V such that $\sigma(a, M) = 0$. If the bilinear form σ is understood we shall often write $\perp$ for $\perp(\sigma)$ and $\top$ for $\top(\sigma)$.

From the definitions we see that $M \subset N$ implies both $M^\perp \supset N^\perp$ and $M^\top \supset N^\top$. Moreover, since σ is bilinear we see that $M^\perp$ and $M^\top$ are always subspaces of V. Clearly also $0^\perp = 0^\top = V$.

The crucial result for us is the following.

PROPOSITION 4.   *If σ is a bilinear form on the vector space V and M is any subspace of V, then*
  (1) $\dim M + \dim M^\perp = \dim V + \dim (M \cap V^\top)$,
  (2) $\dim M + \dim M^\top = \dim V + \dim (M \cap V^\perp)$.

PROOF.   We first give a proof in terms of coordinates. To simplify the argument we may choose an ordered basis $(a_1, \ldots, a_n)$ of V which extends an ordered basis $(a_1, \ldots, a_m)$ of M. Let b have coordinate row $y = (y_1, \ldots, y_n)$ with respect to $(a_1, \ldots, a_n)$ and let A be the $m \times n$ matrix $(\sigma(a_i, a_j))$ $(i = 1, \ldots, m; j = 1, \ldots, n)$. ($A$ consists of the top m rows of the matrix S defined in the previous section.) Now $b \in M^\perp$ if, and only if, $\sigma(a_i, b) = 0$ for $i = 1, \ldots, m$. Thus $b \in M^\perp$ if, and only if, $\sigma(a_i, b) = 0$ for $i = 1, \ldots, m$. Thus $b \in M^\perp$ if, and only if, y is a solution of the linear equation $Ay^t = 0$.

The argument now boils down to finding the row rank of A. In order to do this we go back to our choice of basis for M and insist (as we may) that $(a_1, \ldots, a_m)$ extends an ordered basis $(a_{r+1}, \ldots, a_m)$ of $M \cap V^\top$. By the definition of $V^\top$ this implies that the last $(m-r)$ rows of A are all zero. Suppose that there is a linear relation

$$x_1\sigma(a_1, a_j) + \cdots + x_r\sigma(a_r, a_j) = 0 \quad (j = 1, \ldots, n)$$

among the first r rows. It follows that

$$\sigma(x_1a_1 + \cdots + x_ra_r, a_j) = 0 \quad (j = 1, \ldots, n)$$

from which $x_1a_1 + \cdots + x_ra_r \in V^\mathsf{T}$. By our choice of basis this gives $x_1 = x_2 = \cdots = x_r = 0$ and so the first r rows of A are linearly independent. We may now appeal to Theorem 4 of Chapter III (p. 50) which shows that there are $(n - r)$ linearly independent solutions of the equation $Ay^t = 0$ and so the dimension of $M^\perp$ is $(n - r)$. In other words

$$\dim M^\perp = \dim V - (\dim M - \dim (M \cap V^\mathsf{T}))$$

from which equation (1) follows.

The other equation is proved in the same way.

We shall give another, and in some ways more instructive, proof of Proposition 4. It exploits the link between $\perp$, T and the annihilator mapping o of § 4.6. The fact that there should be such a link is perhaps made most obvious by the observation that $V^\perp = \operatorname{Ker} \tilde{\sigma}$: for $V^\perp$ consists of all b in V such that $a(b\tilde{\sigma}) = 0$ for all a in V. Similarly $V^\mathsf{T} = \operatorname{Ker} g$.

LEMMA 1. $M^{\perp(\sigma)} = (Mg)^\circ$ and $M^{\mathsf{T}(\sigma)} = (M\tilde{\sigma})^\circ$.

PROOF. The vector b lies in $(Mg)^\circ$ if, and only if, $bf = 0$ for all f in Mg, i.e., for all $f = ag$ with a in M. By the definition of g this is equivalent to $\sigma(a, b) = 0$ for all a in M, i.e., to $b \in M^{\perp(\sigma)}$. The second part is proved in exactly the same way.

SECOND PROOF OF PROPOSITION 4. By Lemma 1, $M^\perp = (Mg)^\circ$. Now

$$\dim (Mg)^\circ = \dim V - \dim Mg,$$

by Theorem 6 (1), Chapter IV (p. 81), and

$$\dim Mg = \dim M - \dim (M \cap \operatorname{Ker} g),$$

by Theorem 1, Chapter IV (p. 66). Hence by substitution,

$$\dim M^\perp = \dim V - \dim M + \dim (M \cap V^\mathsf{T}),$$

as required.

If we put $M = V$ in (1) or (2) we have the

COROLLARY. $\dim V^\perp = \dim V^\mathsf{T}$.

Since $\operatorname{rank} g = \dim V - \dim V^\mathsf{T}$ and $\operatorname{rank} \tilde{\sigma} = \dim V - \dim V^\perp$ (by Theorem 1 of Chapter IV (p. 66)), this corollary gives a matrix-free proof of the fact that the ranks of g and $\tilde{\sigma}$ are equal.

Our discussion so far has shown that the mappings $\perp$ and T have very similar properties. In the following lemma and in Theorem 1 we exhibit a direct relationship between them. Theorem 1 also brings out forcefully the connexion between $\perp$, T on the one hand and the annihilator mapping on the other.

LEMMA 2. $M \subset M^{\perp\top}$ and $M \subset M^{\top\perp}$ for every subspace M of V.

PROOF.   If a is any vector in M and b is any vector in $M^{\perp}$, then, by the definition of $M^{\perp}$, $\sigma(a, b) = 0$. But this implies that a is an element of $(M^{\perp})^{\top}$ and thus $M \subset M^{\perp\top}$. The other inclusion is proved similarly.

THEOREM 1.   Let σ be a non-degenerate bilinear form on a vector space V. Then the mappings $\perp(\sigma)$, $\top(\sigma)$ are one–one mappings of the set of all subspaces of V onto itself and have the following properties:

(1) dim $M^{\perp} =$ dim $V -$ dim $M =$ dim $M^{\top}$;

(2) $M^{\perp\top} = M = M^{\top\perp}$;

(3) $M \subset N$ if, and only if, $M^{\perp} \supset N^{\perp}$,
     $M \subset N$ if, and only if, $M^{\top} \supset N^{\top}$;

(4) $(M + N)^{\perp} = M^{\perp} \cap N^{\perp}$,
     $(M + N)^{\top} = M^{\top} \cap N^{\top}$;

(5) $(M \cap N)^{\perp} = M^{\perp} + N^{\perp}$,
     $(M \cap N)^{\top} = M^{\top} + N^{\top}$.

PROOF.   The non-degeneracy of σ implies that $V^{\top} = 0 = V^{\perp}$ and so (1) is an immediate consequence of Proposition 4.

By Lemma 2, $M \subset M^{\perp\top}$ and by (1) dim $M^{\perp\top} =$ dim M, whence (2) holds.

The rest of the proof is entirely analogous to the proof of Theorem 6 of Chapter IV (p. 81).

If σ is a bilinear form on a vector space V and M is a subspace, we shall denote by σ_M the restriction of σ to M.

DEFINITION.   If σ_M is degenerate (or non-degenerate) then we say that M is a *degenerate* (or *non-degenerate*) subspace of V with respect to σ.

LEMMA 3.   *The following three conditions are equivalent:*

(1) M *is a non-degenerate subspace of* V *with respect to* σ;

(2) $M \cap M^{\perp(\sigma)} = 0$;

(3) $M \cap M^{\top(\sigma)} = 0$.

PROOF.   By definition, $M^{\perp(\sigma_M)} = M \cap M^{\perp(\sigma)}$ and $M^{\top(\sigma_M)} = M \cap M^{\top(\sigma)}$ so that conditions (2) and (3) are equivalent by Proposition 4, Corollary. By the remarks following that corollary they are also both equivalent to the condition that σ_M should have maximum rank dim M, i.e., to condition (1).

COROLLARY.   *If* $V = M \oplus V^{\perp}$ *and we assume* $V^{\perp} = V^{\top}$, *then* M *is a non-degenerate subspace of* V *with respect to* σ.

PROOF.   We verify condition (2) of the lemma. If $b \in M^{\perp}$ then $\sigma(M, b) = 0$. But $\sigma(V^{\perp}, b) = 0$ because $V^{\perp} = V^{\top}$, and therefore

$\sigma(M + V^{\perp}, b) = 0$. Hence $b \in V^{\perp}$. We have thus shown that $M^{\perp} \subset V^{\perp}$ which implies that $M \cap M^{\perp} = 0$.

The following result is of basic importance for the structure theorems of § 5.4.

PROPOSITION 5. *Let σ be a bilinear form on the vector space V. If M is a non-degenerate subspace of V with respect to σ then*

$$V = M \oplus M^{\perp} \quad and \quad V = M \oplus M^{\top}.$$

PROOF. We shall confine our attention to $\perp$: the argument for $\top$ is exactly the same. By Lemma 3, $M \cap M^{\perp} = 0$ and so we only have to show that $M + M^{\perp} = V$. For this it suffices to prove dim $M + $ dim $M^{\perp} = $ dim V. Now $M \cap V^{\top} \subset M \cap M^{\top}$ and $M \cap M^{\top} = 0$, by Lemma 3. Hence the required dimension equality follows from Proposition 4.

EXERCISES

1. If S is a subset of V and $S^{\perp}$ is the set of all b in V such that $\sigma(S, b) = 0$, prove that $S^{\perp}$ is a subspace and that $S^{\perp} = [S]^{\perp}$.

2. If $\sigma \in \mathscr{B}(V)$ show that $(\text{Ker } g)^{\perp} = V$ and that $(M^{\top})g = M^{\circ} \cap Vg$.

3. Let V be the vector space F^4 and let σ be the bilinear form on V whose matrix with respect to the standard basis $(e_1, \ldots, e_4)$ is

$$\begin{pmatrix} 1 & 1 & 0 & 0 \\ 0 & 0 & 0 & 0 \\ 1 & 0 & 0 & 0 \\ 0 & 1 & 0 & 0 \end{pmatrix}.$$

   If $M = [e_1, e_2]$, prove that (i) $M \cap M^{\perp} \neq 0$, (ii) $M \cap V^{\perp} = 0$ and (iii) dim $M^{\perp} = 3$. Show further that $V = V^{\perp} \oplus V^{\top}$.

   (Note that (ii) and (iii) together show that Proposition 4 (1) fails if $V^{\perp}$ replaces $V^{\top}$.)

4. If there exists a subspace M of V such that $M^{\perp}$ is a non-degenerate subspace with respect to σ, prove that σ is non-degenerate.

5. Let ρ be the bilinear form on F^n defined by $\rho(x, y) = xy^t$ for all x, y in F^n. We shall refer to ρ as the *standard bilinear form* on F^n.

   Find the matrix of ρ with respect to the standard ordered basis $(e_1, \ldots, e_n)$ of F^n (p. 13). Prove that $\perp(\rho) = \top(\rho)$ and that ρ is non-degenerate.

6. Let x, y be non-zero elements of F^n.
   (i) Use exercise 5 to show that $x^{\perp(\rho)}$ and $y^{\perp(\rho)}$ are hyperplanes in F^n.
   (ii) Show that, if $x^{\perp(\rho)} = y^{\perp(\rho)}$, then $[x] = [y]$.
   Interpret these two results in the language of Theorem 4 of Chapter III (p. 50).

5.3   Symmetric and Alternating Bilinear Forms

As we saw in the last section, orthogonality is not in general a symmetric relation. It is precisely in this case, however, that the concept of orthogonality is of greatest importance.

DEFINITION. A bilinear form σ on a vector space V will be called *orthosymmetric* if

$$\sigma(a, b) = 0 \quad \text{is equivalent to} \quad \sigma(b, a) = 0$$

for all a, b in V.

It is clear that $\perp(\sigma) = \top(\sigma)$ for an orthosymmetric bilinear form. On the other hand if $\perp(\sigma) = \top(\sigma)$, then the equality of $[a]^{\perp}$ and $[a]^{\top}$ shows that $\sigma(a, b) = 0$ if, and only if, $\sigma(b, a) = 0$. Thus the condition of orthosymmetry is equivalent to $\perp(\sigma) = \top(\sigma)$.

If we define $\tau(a, b) = \sigma(b, a)$, then τ is clearly a bilinear form on V and the condition $\perp(\sigma) = \top(\sigma)$ is equivalent to $\perp(\sigma) = \perp(\tau)$.

LEMMA 4. *If σ, τ are bilinear forms on V such that $\perp(\sigma) = \perp(\tau)$, then $\sigma = z\tau$ for some non-zero scalar z.*

PROOF. The condition $\perp(\sigma) = \perp(\tau)$ implies that $[a]^{\perp(\sigma)} = [a]^{\perp(\tau)}$ for all a in V and hence that

$$\sigma(a, b) = 0 \quad \text{if, and only if,} \quad \tau(a, b) = 0 \quad \text{for all } a, b \text{ in } V. \qquad (1)$$

This condition in turn implies that $\top(\sigma) = \top(\tau)$. We may therefore choose an ordered basis $(a_1, \ldots, a_n)$ of V such that $(a_{r+1}, \ldots, a_n)$ is an ordered basis of $V^{\top(\sigma)} = V^{\top(\tau)}$. Now consider the matrices $S = (\sigma(a_i, a_j))$, $T = (\tau(a_i, a_j))$ defined as in § 5.1. In each case the last $(n-r)$ rows consist of zeros. We may also verify that the first r rows of S and T are linearly independent (cf. the first proof of Proposition 4). For, if

$$x_1\sigma(a_1, a_j) + \cdots + x_r\sigma(a_r, a_j) = 0 \quad (j = 1, \ldots, n)$$

then

$$\sigma(x_1 a_1 + \cdots + x_r a_r, a_j) = 0 \quad (j = 1, \ldots, n)$$

would imply that $x_1 a_1 + \cdots + x_r a_r \in V^{\top(\sigma)}$, whence $x_1 = \cdots = x_r = 0$ by our choice of basis for V. The same proof applies to T.

Let us interpret condition (1) in terms of coordinates: it reads

$$x S y^t = 0 \quad \text{if, and only if,} \quad x T y^t = 0 \quad \text{for all } x, y \text{ in } F^n. \qquad (2)$$

Let ρ denote the "standard" bilinear form on F^n defined by

$$\rho(x, y) = xy^t$$

(cf. exercise 5 of § 5.2). Then (2) implies that

$$[x\,S]^{\perp(\rho)} = [x\,T]^{\perp(\rho)} \quad \text{for all } x \text{ in } F^n$$

and hence (exercise 6 of § 5.2) we have

$$[x\,S] = [x\,T] \quad \text{for all } x \text{ in } F^n.$$

In particular, if $(e_1, \ldots, e_n)$ is the standard basis of F^n (p. 13) we have

$$e_1 S = z_1 e_1 T,$$
$$\cdots$$
$$e_r S = z_r e_r T,$$

and

$$(e_1 + \cdots + e_r)S = z(e_1 + \cdots + e_r)T,$$

where $z_1, \ldots, z_r, z$ are non-zero scalars. (Recall that $e_i S$, $e_i T$ are the i-th rows of S, T, respectively.) It follows that

$$(z_1 e_1 + \cdots + z_r e_r)T = z(e_1 + \cdots + e_r)T,$$

and the linear independence of the first r rows of T allows us to deduce that $z_1 = z_2 = \cdots = z_r = z$. Bearing in mind the fact that the remaining rows of S and T are zero, we have $S = zT$ and so $\sigma = z\tau$, as required.

A more illuminating and faster way of proving Lemma 4 is as follows:

SECOND PROOF OF LEMMA 4. By Lemma 1 of the previous section (p. 94), $M^{\perp(\sigma)} = (M\varrho)^\circ$ and $M^{\perp(\tau)} = (M\tau)^\circ$. Thus $M\varrho = M\tau$ for all subspaces M of V. In the language of § 4.2 this means that the (possibly degenerate) projectivities $\mathscr{P}(\varrho)$, $\mathscr{P}(\tau)$ are equal. Let K be the common kernel of ϱ and τ and let M be any subspace such that $M \oplus K = V$. By Theorem 1 of Chapter IV (p. 66), the restrictions of ϱ and τ to M are isomorphisms and so we may apply Proposition 1 of Chapter III (p. 45) to see that $\varrho = z\tau$ for some non-zero scalar z. Hence $\sigma = z\tau$ by Proposition 3.

The lemma shows that if σ is orthosymmetric, then

$$\sigma(a, b) = z\sigma(b, a)$$

for all a, b in V and some non-zero scalar z. This implies that

$$\sigma(a, b) = z\sigma(b, a) = z^2\sigma(a, b)$$

for all a, b in V. If there are any vectors a, b in V such that $\sigma(a, b) \neq 0$ we deduce that $z = 1$ or -1.

DEFINITION. If $\sigma(a, b) = \sigma(b, a)$ for all a, b in V, then σ is called a *symmetric bilinear form* and (V, σ) a *symmetric* (or sometimes an *orthogonal*) *space*.

If $\sigma(a, b) = -\sigma(b, a)$ for all a, b in V, then σ is called a *skew-symmetric bilinear form* and (V, σ) a *skew-symmetric space*.

We have therefore proved

PROPOSITION 6.   *A bilinear form is orthosymmetric if, and only if, it is either symmetric or skew-symmetric.*

Suppose that σ is a skew-symmetric bilinear form.   Then

$$\sigma(a, a) = -\sigma(a, a)$$

for all a in V.   In other words

$$2\sigma(a, a) = 0$$

for all a in V.

If we assume that the ground field is not of characteristic 2 then $2 \neq 0$ (cf. p. 40).   We can therefore find an inverse of 2 in F (axiom F.2, p. 5) and deduce that

$$\sigma(a, a) = 0$$

for all a in V.

On the other hand if F is a field of characteristic 2, then the notions of symmetric and skew-symmetric bilinear forms coincide and we cannot deduce that $\sigma(a, a)$ vanishes.

DEFINITION.   If $\sigma(a, a) = 0$ for all a in V then σ is called an *alternating bilinear form* and (V, σ) an *alternating* (or *symplectic*) *space*.

If σ is alternating, then for any a, b in V

$$0 = \sigma(a+b, a+b) = \sigma(a, a) + \sigma(a, b) + \sigma(b, a) + \sigma(b, b)$$

whence

$$\sigma(a, b) = -\sigma(b, a)$$

i.e., σ is skew-symmetric.

We may summarize as follows:

(1) If the ground field is not of characteristic 2 then an alternating space is exactly the same as a skew-symmetric space.

(2) If the ground field is of characteristic 2 then an alternating space is a special kind of symmetric space.

Suppose now that σ is a symmetric bilinear form on V.   We may define a mapping q of V into F by

$$q(a) = \sigma(a, a)$$

for all a in V.

If $(a_1, \ldots, a_n)$ is an ordered basis of V, $\sigma(a_i, a_j) = s_{ij}$ and $a = \sum x_i a_i$, then, by the bilinearity of σ,

$$q(a) = \sigma(a, a) = \sum s_{ij} x_i x_j.$$

The homogeneous polynomial

$$f(X_1, \ldots, X_n) = \sum s_{ij} X_i X_j$$

is called the *quadratic polynomial of* σ with respect to the ordered basis $(a_1, \ldots, a_n)$. Of course this polynomial is obtained from the (symmetric) bilinear polynomial $\sum s_{ij} X_i Y_j$ by setting $Y_i = X_i$, $i = 1, \ldots, n$.

The function q satisfies the relations

$$q(xa) = x^2 q(a), \tag{3}$$

$$q(a+b) - q(a) - q(b) = 2\sigma(a, b), \tag{4}$$

for all a, b in V.

In studying symmetric bilinear forms in later sections we shall usually make the simplifying assumption that the field F is not of characteristic 2 and then equation (4) shows at once that σ is uniquely determined by q. We can thus refer to $f(X_1, \ldots, X_n)$ as the *quadratic polynomial of* q (with respect to $(a_1, \ldots, a_n)$).

It is, however, interesting to note that when F is of characteristic 2, equation (4) becomes

$$q(a+b) = q(a) + q(b)$$

for all a, b in V. In other words, q is a homomorphism of the additive group V into the additive group F. If we interpret $q(a)$ as a "squared length" we may say in more geometric language that the conclusion of Pythagoras' Theorem is true for *any* triangle in V! In this case the quadratic polynomial of σ with respect to $(a_1, \ldots, a_n)$ is a "sum of squares", i.e.,

$$f(X_1, \ldots, X_n) = \sum s_{ii} X_i^2.$$

Since $s_{ii} = q(a_i)$ it still makes sense to refer to $f(X_1, \ldots, X_n)$ as the quadratic polynomial of q, although here q does not determine the bilinear form σ uniquely.

When there is no restriction on the ground field F, we make the following

DEFINITION. A mapping q of V into F is called a *quadratic form* on V if

Q.1. $q(xa) = x^2 q(a)$,

for all a in V and x in F, and

Q.2. $q(a+b) - q(a) - q(b) = \varphi(a, b)$

defines a bilinear form φ on V.

Clearly such a φ is symmetric and moreover $\varphi(a, a) = 2q(a)$. Hence if F is not of characteristic 2, $q(a) = \frac{1}{2}\varphi(a, a)$, i.e., q is the "squared length" in the symmetric space $(V, \frac{1}{2}\varphi)$. But if F is of characteristic 2, then φ is both symmetric and alternating and we regard our space as a refinement of an alternating space (V, φ) in which a squared length is also given. (The interested reader should consult [5]; also exercises 8, 9 below.)

EXERCISES

1. Prove that a bilinear form is symmetric or skew-symmetric if, and only if, its matrix with respect to any one ordered basis is symmetric or skew-symmetric. (Recall that a matrix A is symmetric if $A = A^t$; it is *skew-symmetric* if $A = -A^t$.)

2. Prove that a bilinear form σ is symmetric if, and only if, $\varrho = \bar{\sigma}$.

3. Show that σ is orthosymmetric if, and only if, $M \subset M^{\perp\perp}$ for all subspaces M.

4. If $\mathscr{B}^+(V)$ denotes the set of all symmetric bilinear forms on V and $\mathscr{B}^-(V)$ the set of all skew-symmetric ones, prove that $\mathscr{B}^+(V)$ and $\mathscr{B}^-(V)$ are subspaces of $\mathscr{B}(V)$ and if F is not of characteristic 2 then

$$\mathscr{B}(V) = \mathscr{B}^+(V) \oplus \mathscr{B}^-(V).$$

5. Prove that $q: x \to x\bar{x}$ is a quadratic form on $\mathbf{C}$, viewed as a two-dimensional real vector space (where the bar denotes complex conjugation). Show further that, when q is restricted to the multiplicative group $\mathbf{C}^*$, it is a group homomorphism of $\mathbf{C}^*$ into $\mathbf{R}^*$. What is the kernel and what is the image?

6. If D is the real quaternion algebra as given in exercise 7 of § 4.3 (p. 75), prove that

$$q: \begin{pmatrix} x & y \\ -\bar{y} & \bar{x} \end{pmatrix} \to x\bar{x} + y\bar{y}$$

is a quadratic form on D and that q restricted to D^* is a group homomorphism of D^* into $\mathbf{R}^*$. Find the quadratic polynomial of q with respect to the ordered basis

$$\left(\begin{pmatrix} 1 & 0 \\ 0 & 1 \end{pmatrix}, \begin{pmatrix} i & 0 \\ 0 & -i \end{pmatrix}, \begin{pmatrix} 0 & 1 \\ -1 & 0 \end{pmatrix}, \begin{pmatrix} 0 & i \\ i & 0 \end{pmatrix} \right)$$

of D.

7. Let V be a vector space over $\mathbf{F}_2$ and q a quadratic form on V with associated bilinear form φ. Prove that q is a linear form on $V^{\perp(\varphi)}$.

8. Let $(a_1, \ldots, a_n)$ be an ordered basis of V and q a quadratic form on V. We define the matrix of q with respect to $(a_1, \ldots, a_n)$ to be (q_{ij}), where

$$q_{ii} = q(a_i),$$
$$q_{ij} = \varphi(a_i, a_j) \quad \text{if} \quad i \neq j,$$

where φ is the bilinear form associated with q. Prove that

$$q(x_1 a_1 + \cdots + x_n a_n) = \sum_{i \leq j} q_{ij} x_i x_j.$$

(Use induction on n. Observe that the result is valid irrespective of whether F is of characteristic 2 or not.)

9. Let $\mathscr{Q}(V)$ be the set of all quadratic forms on V. If q, $q' \in \mathscr{Q}(V)$ and $x \in F$, verify that the mappings

$$q + q' : \ a \to q(a) + q'(a),$$
$$xq : \ a \to x\,q(a)$$

are both quadratic forms on V and that $\mathscr{Q}(V)$ is a vector space over F with respect to this addition and multiplication by scalars.

Using also the notation of exercise 4, let α be the mapping $\sigma \to q$ of $\mathscr{B}^+(V)$ into $\mathscr{Q}(V)$ defined by $q(a) = \sigma(a, a)$ for all a in V; and let β be the mapping $q \to \varphi$ of $\mathscr{Q}(V)$ into $\mathscr{B}^+(V)$ defined by $\varphi(a, b) = q(a+b) - q(a) - q(b)$ for all a, b in V. Show that (1) α, β are linear mappings; (2) $\alpha\beta = 2(1_{\mathscr{B}^+})$, $\beta\alpha = 2(1_{\mathscr{Q}})$; (3) if F is not of characteristic 2 then α, β are isomorphisms; (4) if F is of characteristic 2 then Ker α = Image β and Ker β = Image α.

5.4   Structure Theorems

We saw in § 5.2 that a non-degenerate subspace M of V has $M^\perp$ as direct complement, i.e., $V = M \oplus M^\perp$. In this section we shall use this fundamental result to gain information about the structure of alternating and symmetric spaces.

THEOREM 2. *If (V, σ) is an alternating space, then*

$$V = L_1 \oplus \cdots \oplus L_s \oplus V^\perp,$$

where $L_1, \ldots, L_s$ are mutually orthogonal non-degenerate two-dimensional subspaces and $s = \frac{1}{2}$ (rank σ).

PROOF. If $\sigma(b, c) = 0$ for all b and c, then $V = V^\perp$ and we are finished. So suppose $\sigma(b, c) \neq 0$ for some b, c in V. Then b, c must be linearly independent, and we assert that $L = [b, c]$ is non-degenerate: for if $v = xb + yc$ lies in $L^\perp$, then $\sigma(b, v) = 0$ and $\sigma(c, v) = 0$ yield $y = x = 0$. Hence, by Proposition 5, $V = L \oplus L^\perp$ and the direct sum decomposition of V follows by induction on dim V. Finally, $2s = \dim V - \dim V^\perp$ and thus $2s = \text{rank } \sigma$ by the corollary to Proposition 4.

COROLLARY 1. *A non-degenerate alternating space must have even dimension.*

Each subspace L_i in Theorem 2 has an ordered basis (b_i, c_i) such that $\sigma(b_i, c_i) = x_i \neq 0$. Then with respect to $(b_i, (1/x_i) c_i)$, the restriction of σ to L_i has matrix

$$\begin{pmatrix} 0 & 1 \\ -1 & 0 \end{pmatrix}.$$

Hence Theorem 2 yields

COROLLARY 2. *There exists an ordered basis with respect to which σ has matrix*

$$A_{n,\,s} = \begin{pmatrix} 0 & 1 & & & & & & & & \\ -1 & 0 & & & & & & & & \\ & & 0 & 1 & & & & & & \\ & & -1 & 0 & & & & & & \\ & & & & \ddots & & & & & \\ & & & & & 0 & 1 & & & \\ & & & & & -1 & 0 & & & \\ & & & & & & & 0 & & \\ & & & & & & & & 0 & \\ & & & & & & & & & 0 \\ & & & & & & & & & & \ddots \\ & & & & & & & & & & & 0 \end{pmatrix}$$

where n is the number of rows and s is the number of blocks

$$\begin{pmatrix} 0 & 1 \\ -1 & 0 \end{pmatrix}.$$

It is easily seen that a matrix $A = (a_{ij})$ is the matrix of an alternating bilinear form if, and only if, $a_{ij} = -a_{ji}$ whenever $i \neq j$ and $a_{ii} = 0$ for all i. We call such a matrix an *alternating matrix*. Corollary 2 shows that every $n \times n$ alternating matrix of rank $2s$ is congruent to the matrix $A_{n,s}$. This solves the congruence problem for alternating matrices for any ground field (cf. p. 86).

COROLLARY 3. *Two alternating bilinear forms on the same space are congruent if, and only if, they have the same rank.*

We now turn to the structure of symmetric spaces.

THEOREM 3. *Let σ be a symmetric, but not alternating, bilinear form on the vector space V. Then*

$$V = A_1 \oplus \cdots \oplus A_r \oplus V^\perp$$

where $A_1, \ldots, A_r$ are non-degenerate mutually orthogonal one-dimensional subspaces. Moreover $r = \text{rank } \sigma$.

The reader should observe that if F is not of characteristic 2 then an alternating symmetric bilinear form is zero. Thus the condition that σ should not be alternating is only relevant to the special case where F is of characteristic 2.

PROOF. The last assertion of Theorem 3 follows from the corollary to Proposition 4. The direct sum decomposition will be proved by an induction on dim V.

Since σ is not alternating there must exist a in V such that $\sigma(a, a) \neq 0$. Then $A = [a]$ is a non-degenerate subspace and hence $V = A \oplus A^\perp$ by Proposition 5. If the restriction of σ to $A^\perp$ is either zero or is not alternating the theorem follows by induction. (By the remark above, the proof ends here if F is not of characteristic 2.)

Suppose then that the restriction of σ to $A^\perp$ is alternating and not zero. Thus we can find vectors b, c in $A^\perp$ such that $\sigma(b, c) \neq 0$. We note that $\sigma(a + b, a + b) = \sigma(a, a) \neq 0$ and so $[a + b]$ is non-degenerate. Consider the vector $v = xa + c$. Then

$$\sigma(v, v) = x^2 \sigma(a, a)$$

and

$$\sigma(a + b, v) = x\sigma(a, a) + \sigma(b, c).$$

If $x = -\sigma(b, c)/\sigma(a, a)$, the vector v lies in $[a + b]^\perp$ and $\sigma(v, v) \neq 0$. Thus $V = [a + b] \oplus [a + b]^\perp$ where $[a + b]$ is non-degenerate and σ is not alternating on $[a + b]^\perp$. The proof can now be completed by induction.

Theorem 3 has a corollary for matrices. If $A = (a_{ij})$ is a square matrix, then A is called *diagonal* if $a_{ij} = 0$ whenever $i \neq j$. We shall later use the notation

$$A = \text{diag}(a_{11}, \ldots, a_{nn})$$

when A is diagonal.

COROLLARY. *A symmetric matrix which is not alternating is congruent to a diagonal matrix.*

PROOF. Let $V = F^n$ and let σ be the bilinear form on V such that

$$(\sigma; (e_i)) = A.$$

By Theorem 3 we can find vectors $a_1, \ldots, a_r$ such that

$$V = [a_1] \oplus \cdots \oplus [a_r] \oplus V^\perp$$

with $\sigma(a_i, a_j) = 0$ for $i \neq j$. If $a_{r+1}, \ldots, a_n$ are chosen to form a basis of $V^\perp$, then

$$B = (\sigma; (a_1, \ldots, a_n))$$

is diagonal.

The congruence problem for symmetric matrices is still unsolved except for certain special fields. The above corollary reduces the problem to the case of diagonal matrices, but the final solution depends essentially on the structure of the underlying field. However, the classification of the symmetric bilinear forms over the complex numbers and over the real numbers is easy and will be given below. Moreover, the reader will find the classification over finite fields in exercise 11.

PROPOSITION 7. *If σ is a symmetric bilinear form on a complex vector space V, then there exists an ordered basis of V with respect to which σ has matrix*

$$\begin{pmatrix} I_r & 0 \\ 0 & 0 \end{pmatrix},$$

where r is the rank of σ.

PROOF. By the corollary to Theorem 3, there exists an ordered basis $(a_1, \ldots, a_n)$ such that

$$(\sigma; (a_i)) = \text{diag}(x_1, \ldots, x_r, 0, \ldots, 0),$$

with $x_i \neq 0$, $i = 1, \ldots, r$. Let $y_i^2 = x_i^{-1}$, $i = 1, \ldots, r$. Then

$$(\sigma; (y_1 a_1, \ldots, y_r a_r, a_{r+1}, \ldots, a_n)) = \begin{pmatrix} I_r & 0 \\ 0 & 0 \end{pmatrix}.$$

The reader should note that the only property of C we use in the proof is that every complex number has at least one square root.

COROLLARY. *Two symmetric bilinear forms on a complex vector space, or two symmetric matrices in $C^{n \times n}$, are congruent if, and only if, they have the same rank.*

Now let (V, σ) be an n-dimensional real symmetric space and suppose that $(a_1, \ldots, a_n)$ is an ordered basis for which

$$(\sigma; (a_i)) = \text{diag}(x_1, \ldots, x_r, 0, \ldots, 0).$$

We may assume $(a_1, \ldots, a_n)$ so ordered that $x_1, \ldots, x_p$ are positive and $x_{p+1}, \ldots, x_{p+q}$ are negative, where $p + q = r$. If we replace $a_1, \ldots, a_p$ by

$$\frac{1}{\sqrt{x_1}} a_1, \ldots, \frac{1}{\sqrt{x_p}} a_p$$

and $a_{p+1}, \ldots, a_{p+q}$ by

$$\frac{1}{\sqrt{-x_{p+1}}} a_{p+1}, \ldots, \frac{1}{\sqrt{-x_{p+q}}} a_{p+q},$$

then the matrix of σ with respect to the new basis is

$$\text{diag} \, (\underbrace{1, \ldots, 1}_{p}, \underbrace{-1, \ldots, -1}_{q}, 0, \ldots, 0)$$

In order to prove the theorem of Sylvester that the integers p and q so defined are invariants of (V, σ) we shall introduce a new notion.

DEFINITION. A real-valued symmetric bilinear form σ is *positive semi-definite* if $q(a) = \sigma(a, a) \geq 0$ for all a. If further $q(a) = 0$ implies $a = 0$ then σ is *positive definite*. In these circumstances we shall also say that the quadratic form q is positive semi-definite or positive definite. The definition of *negative semi-definite* and *negative definite* bilinear forms (and quadratic forms) is similar.

PROPOSITION 8. *Let (V, σ) be a real symmetric space. If*

$$V = P \oplus Q \oplus V^\perp$$

where the restriction of σ to P is positive definite and the restriction of σ to Q is negative definite, then $\dim P$ and $\dim Q$ are invariants of (V, σ).

PROOF. If M is a positive definite subspace and N is a negative semi-definite subspace then clearly $M \cap N = 0$. Thus

$$\dim M + \dim N \leq \dim V.$$

But in particular $N = Q \oplus V^\perp$ satisfies this inequality and so

$$\dim M \leq \dim V - (\dim V - \dim P) = \dim P.$$

Hence $p = \dim P$ is the maximum dimension of all the positive definite subspaces of V, and as such is an invariant. Finally, $q = \dim Q$ is also an invariant because $q = r - p$ where r is the rank of σ.

DEFINITION. $p - q$ is the *signature* of σ.

PROPOSITION 9. *If σ is a symmetric bilinear form on a real vector space V, then there exists an ordered basis of V with respect to which σ has matrix*

$$\begin{pmatrix} I_p & 0 & 0 \\ 0 & -I_q & 0 \\ 0 & 0 & 0 \end{pmatrix},$$

where $p + q$ is the rank of σ and $p - q$ is the signature of σ.

COROLLARY. (SYLVESTER'S "LAW OF INERTIA".) *Two symmetric bilinear forms on a real vector space are congruent if, and only if, they have the same rank and same signature.*

Sylvester's Law implies that there are essentially (i.e., to within isomorphism) $n+1$ distinct non-degenerate real symmetric spaces of dimension n. We remark that the "space time" of relativity has rank 4 and signature 2.

The next chapter will be devoted to the study of the positive definite spaces—the so called *euclidean spaces*.

So far in this section we have only considered orthosymmetric bilinear forms. There is a generalization of Theorem 3 which is true for arbitrary non-alternating bilinear forms provided only that we assume that the ground field has more than two elements.

THEOREM 4. *Let σ be a non-alternating bilinear form on a vector space V over a field F containing more than two elements. Then*

$$V = A_1 \oplus \cdots \oplus A_r \oplus V^\perp$$

where $A_1, \ldots, A_r$ are non-degenerate one-dimensional subspaces and

$$\sigma(A_i, A_j) = 0 \quad \text{for all} \quad 1 \leq i < j \leq r.$$

Moreover $r =$ rank σ.

PROOF. As in Theorem 3 we argue by induction on dim V.

Because σ is not alternating, there must exist a vector a in V such that $\sigma(a, a) \neq 0$. By Proposition 5, $V = [a] \oplus [a]^\perp$. If the restriction of σ to $[a]^\perp$ is either zero or is not alternating, the theorem follows by induction.

Suppose then that the restriction of σ to $[a]^\perp$ is alternating and not zero. Our task, as before, is to find a vector b in V so that $[a+b]$ is non-degenerate and σ is not alternating on $[a+b]^\perp$. We may find vectors b, c in $[a]^\perp$ such that $\sigma(b, c) \neq 0$. Let $v = xa + c$. We try to satisfy the following three conditions:

    (i) $\sigma(a+b, a+b) = \sigma(a, a) + \sigma(b, a) \neq 0$,

    (ii) $\sigma(a+b, v) = x(\sigma(a, a) + \sigma(b, a)) + \sigma(b, c) = 0$,

    (iii) $\sigma(v, v) = x^2\sigma(a, a) + x\sigma(c, a) \neq 0$.

Let us now assume that $\sigma(b, a) = 0$. Then condition (i) is automatically satisfied, (ii) is satisfied by taking $x = -\sigma(b, c)/\sigma(a, a)$ and (iii) is then equivalent to $\sigma(b, c) \neq \sigma(c, a)$. Since F contains more than two elements, we may satisfy all three conditions in this case by replacing b, if necessary, by a suitable scalar multiple zb.

If $\sigma(b, a) \neq 0$ but $\sigma(c, a) = 0$ we may interchange the roles of b and c (since $\sigma(c, b) = -\sigma(b, c) \neq 0$) and prove the result in the same way.

Finally, let us assume neither $\sigma(b, a)$ nor $\sigma(c, a)$ is zero. If $b' = b + yc$, then

$$\sigma(b', c) = \sigma(b, c) \neq 0$$

and

$$\sigma(b', a) = \sigma(b, a) + y\sigma(c, a).$$

We choose $y = -\sigma(b, a)/\sigma(c, a)$ so that $\sigma(b', a) = 0$. Then we may argue as above with b replaced by b'. This completes the proof.

Theorem 4 has a natural interpretation in terms of matrices.

DEFINITION. If $A = (a_{ij})$ is a square matrix, then A is called (lower) *triangular* if $a_{ij} = 0$ whenever $i < j$ (i.e., if all terms above the diagonal are zero).

COROLLARY. *If A is a non-alternating matrix in $F^{n \times n}$, where F contains more than two elements, then A is congruent to a (lower) triangular matrix.*

Theorem 4 is false when $F = \mathbf{F}_2$: cf. exercise 7 below.

EXERCISES

1. Let (V, σ) be a non-degenerate two-dimensional symmetric space over a field which is not of characteristic 2. If there is a non-zero vector a in V such that $\sigma(a, a) = 0$, prove that there exists a unique b in V such that $\sigma(b, b) = 0$ and $\sigma(a, b) = 1$. Is the condition on the field necessary?

2. Prove that the quadratic forms of exercises 5, 6 of § 5.3 (p.101) are both positive definite.

3. If σ is positive semi-definite and $\sigma(c, c) = 0$ show that $\sigma(a, c) = 0$ for all a.

4. Define the signature of a real symmetric matrix and give a matrix interpretation of the corollary to Proposition 9.

5. All polynomials in this exercise have real coefficients.
   (i) If

$$f(X_1, \ldots, X_n) = \sum_{i, j = 1}^{n} a_{ij} X_i X_j$$

where $a_{ij} = a_{ji}$ for all i, j and $a_{11} \neq 0$, show that the transformation defined by

$$a_{11} Y_1 = a_{11} X_1 + a_{12} X_2 + \cdots + a_{1n} X_n,$$
$$Y_i = X_i, \quad i = 2, \ldots, n$$

leads to an equation

$$f(X_1, \ldots, X_n) = a_{11} Y_1^2 + f'(Y_2, \ldots, Y_n),$$

where f' is another quadratic polynomial.

(ii) If $X_1 = Y_1 - Y_2$ and $X_2 = Y_1 + Y_2$ show that $X_1 X_2 = Y_1^2 - Y_2^2$ and solve for Y_1, Y_2 in terms of X_1, X_2.

(iii) Use steps of the above kind to reduce the following quadratic polynomials to sums of squares and find the rank and signature in each case.

(1) $X_1 X_2 + X_1 X_3 + X_2 X_3$,

(2) $X_1^2 + 4X_1 X_2 + 3X_2^2 + 2X_2 X_3 - X_3^2$,

(3) $X_1^2 - 6X_1 X_2 + 4X_1 X_3 + 9X_2^2 - 9X_2 X_3 + 4X_3^2$.

6. If q is a quadratic form on the vector space V over a field of characteristic 2, prove that V has a basis $\{a_1, \ldots, a_n\}$ such that

$$q(x_1 a_1 + \cdots + x_n a_n) = \left(\sum_{i=1}^{n} q(a_i) x_i^2 \right) + x_1 x_2 + x_3 x_4 + \cdots + x_{2s-1} x_{2s}.$$

(Use exercise 8 of § 5.3 and Theorem 2.)

7. Show that the matrix

$$\begin{pmatrix} 1 & 0 & 0 \\ 1 & 0 & 1 \\ 0 & 1 & 0 \end{pmatrix}$$

is *not* congruent to a triangular matrix if the ground field is F_2.

8. If F is a finite field, show that F contains a field F_0 isomorphic to F_p for some prime p. Prove also that every field E contained in F must contain p^r elements for some r. (Cf. also exercise 10 of § 7.3, p. 162.)

9. Let F be a finite field and F^* the multiplicative group of non-zero elements. Show that $x \to x^2$ is a homomorphism of F^* into itself with kernel $\{+1, -1\}$. Denote the image by S^*. Prove that $S^* = F^*$ if, and only if, F is of characteristic 2.

If F is not of characteristic 2 and z is any element not in S^*, prove that every element of F^* either lies in S^* or has the form zs for some s in S^*.

10. Let F be a finite field not of characteristic 2 and suppose S^* and z have the same meaning as in exercise 9. Assume that z cannot be written as a sum of two squares. Prove that $S = \{0\} \cup S^*$, the set of all squares in F, is a field (using the last part of exercise 9) and that the number of elements in S is $\frac{1}{2}(q+1)$, where q is the number of elements in F. Deduce (using exercise 8) that this is impossible.

(Exercises 9 and 10 yield the result that *every element in a finite field can be written as a sum of two squares*.)

11. Let σ be a symmetric, but not alternating, bilinear form of rank r on the vector space V over the *finite* field F. Using Theorem 3 and the result stated immediately before this exercise, show that there exists an ordered basis of V with respect to which σ has matrix

$$\begin{pmatrix} I_r & 0 \\ 0 & 0 \end{pmatrix} \quad \text{or} \quad \begin{pmatrix} I_{r-1} & 0 & 0 \\ 0 & z & 0 \\ 0 & 0 & 0 \end{pmatrix},$$

where z is a non-square in F.

5.5 Correlations

If σ is a non-degenerate bilinear form on V then we deduce from Lemma 1 (p. 94) that $\perp(\sigma) = \mathscr{P}(g)\circ$ and $\top(\sigma) = \mathscr{P}(\tilde{\sigma})\circ$. In other words, $\perp(\sigma)$ and $\top(\sigma)$ are *correlations* of $\mathscr{P}(V)$. This fact summarises (1), (3), (4), (5) of Theorem 1 on page 95 (and (2) simply means that $\perp^{-1} = \top$ and $\top^{-1} = \perp$). Since the mappings $\sigma \to g$, $\sigma \to \tilde{\sigma}$ of Proposition 3 (p. 91) are onto $\mathscr{L}(V, V^*)$, *any* correlation of $\mathscr{P}(V)$ is of the form $\perp(\sigma)$ (and also of the form $\top(\sigma)$).

In view of Lemma 4 (p. 97) we may define a *polarity* to be a correlation $\perp(\sigma)$ for which σ is symmetric and a *null-polarity* to be a correlation $\perp(\sigma)$ for which σ is alternating. In either case it is usual to refer to $M^\perp$ as the *polar* of M with respect to $\perp$.

There are no null-polarities in even dimensional projective geometries and essentially only one in each odd dimensional projective geometry (Theorem 2, Corollaries 1 and 3 (p. 103)). The geometry of polarities is, by contrast, considerably richer as we shall see in the next section.

If f is an element of $\mathscr{L}(V, V^*)$ with non-zero kernel, then $\mathscr{P}(f)$ is a degenerate projectivity in the sense of § 4.2 and we refer to $\mathscr{P}(f)\circ$ as a *degenerate correlation*. Any degenerate correlation is of the form $\perp(\sigma)$ (or $\top(\sigma)$), where σ is a degenerate bilinear form, and will be called a *degenerate polarity* if σ is symmetric or a *degenerate null-polarity* if σ is alternating.

EXERCISES

1. Let $\perp$ be a (possibly degenerate) correlation. Show that $\perp$ is a polarity or a null-polarity if, and only if, $M \subset M^{\perp\perp}$ for every subspace M; and that $\perp$ is a null-polarity if, and only if, $P \subset P^\perp$ for every point P.

2. Describe the degenerate null-polarities of projective geometries of dimensions 2 and 3.

3. Let $\perp$ be a null-polarity of a projective geometry of dimension 3. If P_1, P_2, P_3, P_4 are the vertices of a tetrahedron T show that the planes $P_1^\perp$, $P_2^\perp$, $P_3^\perp$, $P_4^\perp$ are the faces of a tetrahedron which is both inscribed and circumscribed to T.

4. Let $\perp$ be a null-polarity of a projective geometry of dimension 3. If P, Q are given distinct points such that $P \subset Q^\perp$, prove that the line L joining P and Q satisfies the relation $L = L^\perp$. What is the configuration of all lines through a given point P with this property?

   What is the configuration of all lines L such that $L \subset L^\perp$, when $\perp$ is a degenerate null-polarity?

   (For any null-polarity, degenerate or otherwise, the set of all lines L such that $L \subset L^\perp$ is called a *linear complex*.)

5. Let A be the affine geometry in $\mathscr{P}(V)$ determined by a hyperplane H and let $\perp$ be a polarity of $\mathscr{P}(V)$. We consider the subset B of A obtained by omitting all elements of A which contain the point $H^{\perp}$. Let $\perp'$ be the restriction of $\perp$ to B. We call $\perp'$ the *affine polarity* (of A) *determined by* $\perp$.

    Prove that $\perp'$ is a one–one mapping of B onto B and that $\perp$ is the only polarity of $\mathscr{P}(V)$ which determines $\perp'$.

    (We call $H^{\perp}$ the *center* of $\perp'$. Note that the center need not lie in A.)

6. Let A be the affine geometry in $\mathscr{P}(V)$ determined by a hyperplane H and let $\perp$ be a degenerate polarity of $\mathscr{P}(V)$. We consider the subset B of A obtained by omitting the elements M for which $M^{\perp} \subset H$. Show that $B =$ A if, and only if, $V^{\perp} \in$ A and that B is empty if, and only if, $V^{\perp} = H$.

    When B is not empty we denote by $\perp'$ the restriction of $\perp$ to B and call $\perp'$ the *degenerate affine polarity* (of A) *determined by* $\perp$. Prove that $\perp'$ is a mapping of B into A and that $\perp$ is the only degenerate polarity of $\mathscr{P}(V)$ which determines $\perp'$.

5.6   Projective Quadrics

All vector spaces in this section are over fields which are not of characteristic 2.

DEFINITION. If σ is a symmetric bilinear form on a vector space V then the set of all points $[a]$ of $\mathscr{P}(V)$ for which $\sigma(a, a) = 0$ is a (*projective*) *quadric* and is denoted by $Q(\sigma)$. When pdim $V = 2$ a quadric in $\mathscr{P}(V)$ is called a *conic*.

Since $Q(\sigma) = Q(z\sigma)$ for any non-zero scalar z it is clear from Lemma 4 (p. 97) that there is a mapping $\perp(\sigma) \to Q(\sigma)$. In fact, if $\perp(\sigma)$ is a given polarity (possibly degenerate) then $Q(\sigma)$ is the set of all points P of $\mathscr{P}(V)$ for which $P \subset P^{\perp(\sigma)}$.

LEMMA 5. *The image of a projective quadric under a projectivity is again a projective quadric.*

PROOF. Let g be a linear isomorphism of V onto V'. If σ is a symmetric bilinear form on V then the equation

$$\sigma'(ag, bg) = \sigma(a, b)$$

for all a, b in V, defines a symmetric bilinear form σ' on V'. Then $\sigma(a, a) = 0$ if, and only if, $\sigma'(ag, ag) = 0$. In other words $[a] \in Q(\sigma)$ if, and only if, $[a]\mathscr{P}(g) \in Q(\sigma')$.

Let π be a projective coordinate system for $\mathscr{P}(V)$ and $(a_0, \ldots, a_n)$ an ordered basis of V which determines π (as on page 46). Suppose $f(X_0, \ldots, X_n)$ is the quadratic polynomial of σ with respect to

$(a_0, \ldots, a_n)$. Then $Q(\sigma)$ is the set of all points whose projective coordinates with respect to π satisfy the equation

$$f(X_0, \ldots, X_n) = 0.$$

This is called the *equation of $Q(\sigma)$ with respect to π*. (By Lemma 1 of Chapter III, $f(X_0, \ldots, X_n)$ is determined by σ and π to within a non-zero scalar multiple.) If $S = (\sigma; (a_i))$, then the matrix form of the equation is

$$X \, S \, X^t = 0$$

where $X = (X_0, \ldots, X_n)$. Note that, by Lemma 5, the set $Q(\sigma)\pi$ is a quadric in $\mathscr{P}(F^{n+1})$. Its equation with respect to the standard co-ordinate system coincides with the equation of $Q(\sigma)$ with respect to π.

By Theorem 3, Corollary, there always exists a projective coordinate system with respect to which $Q(\sigma)$ has equation

$$d_0 X_0{}^2 + \cdots + d_r X_r{}^2 = 0,$$

where $d_0 \cdots d_r \neq 0$. (If $\sigma = 0$ this equation reads $0 = 0$.)

Consider, for example, the quadrics on a projective line. If σ has rank 2 then $Q(\sigma)$ is either empty or consists of two distinct points. (Here we use the assumption that the field F is not of characteristic 2.) If σ has rank 1 then $Q(\sigma)$ consists of a single point and if σ has rank 0 then $Q(\sigma)$ consists of all the points on the line.

In the special case of the projective line we can recognize the rank of σ from the set of points $Q(\sigma)$, even if $Q(\sigma)$ is empty. More generally, if $Q(\sigma)$ is a quadric in $\mathscr{P}(V)$ where pdim $V \geq 2$ we can recognize $V^\perp$ as follows. If L is any line in $\mathscr{P}(V)$ then the common points of $Q(\sigma)$ and $\mathscr{P}(L)$ form the quadric $Q(\sigma_L)$. Now the point $[a]$ lies on $V^\perp$ if, and only if, $[a]$ lies on $Q(\sigma)$ and σ_L has rank less than 2 for every line L through $[a]$. (Because of this behavior the points of $V^\perp$ are often called the "singular" points of $Q(\sigma)$ and given the "multiplicity" 2.) We can therefore define the *rank* of $Q(\sigma)$ unambiguously to be the rank of σ (i.e., dim V − dim $V^\perp$) and say that $Q(\sigma)$ is *degenerate* if, and only if, σ is degenerate.

In view of the fact that a quadric may be empty we must not expect always to be able to recapture $\perp (\sigma)$ from $Q(\sigma)$. It can be shown that if $Q(\sigma)$ contains at least *one* point outside $V^\perp$ (e.g., if $Q(\sigma)$ is a non-empty non-degenerate quadric) then $\perp (\sigma)$ is the only polarity determining $Q(\sigma)$. (See exercises 3, 7, 16 where the outlines of a proof are given.)

It is possible to meet this difficulty in another way if one is prepared to extend the ground field and make a corresponding extension of the bilinear form σ. We shall sketch this for the case of a real space. Let $V_{(C)}$ be the complexification of the real vector space V (p. 14) and σ a symmetric bilinear form on V. The extended form σ' must be bilinear over $\mathbf{C}$ and so is given by

$$\sigma'(a+ib, c+id) = \sigma(a, c) - \sigma(b, d) + i(\sigma(a, d) + \sigma(b, c))$$

for all a, b, c, d in V.

We choose an ordered basis $(a_0, \ldots, a_n)$ of V so that $(\sigma; (a_i))$ is diagonal. Now $(a_0, \ldots, a_n)$ is also an ordered basis of $V_{(\mathbf{C})}$ and

$$(\sigma'; (a_i)) = (\sigma; (a_i))$$

by definition. Thus $Q(\sigma)$ and the extended quadric $Q(\sigma')$ have the same equation

$$d_0 X_0{}^2 + \cdots + d_r X_r{}^2 = 0.$$

Clearly $Q(\sigma')$ contains all the points for which $X_p = \sqrt{d_q}$ and $X_q = \pm i\sqrt{d_p}$ where $0 \le p < q \le r$, $X_s = 0$ for $0 \le s \le r$, $s \ne p$, q and X_t is arbitrary for $t > r$. Conversely, any (homogeneous) quadratic polynomial $f(X_0, \ldots, X_n)$ which vanishes for all such points is easily seen to be a scalar multiple of $d_0 X_0{}^2 + \cdots + d_r X_r{}^2$. In other words $Q(\sigma')$ determines $\perp(\sigma')$ uniquely.

We end this section with a simple result which shows the essentially geometrical character of a quadric.

PROPOSITION 10. *The image of a projective quadric under a projective isomorphism is a projective quadric of the same rank.*

PROOF. Let π be a projective isomorphism of $\mathscr{P}(V)$ onto $\mathscr{P}(V')$. The special cases where pdim V equals 0 or 1 are obvious and so we may assume by Theorem 6 of Chapter III (p. 57) that $\pi = \mathscr{P}(g)$ where g is a semi-linear isomorphism of V onto V' with respect to ζ. If σ is a symmetric bilinear form on V then the equation

$$\sigma'(ag, bg) = \sigma(a, b)\zeta$$

for all a, b in V, defines a symmetric bilinear form σ' on V' of the same rank as σ. Just as in Lemma 5 we see that π maps $Q(\sigma)$ onto $Q(\sigma')$.

EXERCISES

1. Show that the equation $aX_0{}^2 + 2h\, X_0 X_1 + bX_1{}^2 = 0$ defines an empty quadric in $\mathscr{P}(\mathbf{R}^2)$ if, and only if, $h^2 < ab$.

2. Describe the degenerate polarity of a quadric of rank 1 on a line L.
   If the quadric $Q(\sigma)$ on L consists of a pair of distinct points A, B show that $(A, B; P, P^{\perp(\sigma)})$ is a harmonic range for every point P on L.

3. Show how a (possibly degenerate) polarity on a line can be reconstructed from its quadric provided the quadric is not empty. (Use exercise 2.)

4. Describe the four types of degenerate conic in $\mathscr{P}(\mathbf{R}^3)$. Show that in three of these cases the degenerate polarity can be reconstructed from the conic.

5. If Q is a non-degenerate conic in a projective plane P and if A is a point on Q, show that every line of P through A, other than the polar line $A^\perp$, intersects Q in two points.

Deduce that there is a one–one mapping of Q onto the points of any line in P.

6. Show that all the points of $\mathscr{P}(\mathbf{Q}^3)$ on the conic with equation

$$X_0{}^2 + X_1{}^2 - 2X_2{}^2 = 0$$

are of the form

$$[(m^2 + 2mn - n^2,\ -m^2 + 2mn + n^2,\ -m^2 - n^2)]$$

where m, n are integers. (Use exercise 5 and note that the point $(1, 1, 1)$ lies on the conic.)

7. Let $Q(\sigma)$ be a non-empty non-degenerate conic. If P is a point not on Q show (by means of exercise 5) that there are at least two lines through P which have points of intersection with Q. Use exercise 3 to give a construction for the polar line $P^{\perp(\sigma)}$ of P.

Give a construction for $P^{\perp(\sigma)}$ when P lies on Q.

8. Let $Q(\sigma)$ be a non-degenerate conic containing three distinct points A_0, A_2, E and let A_1 be the *pole* of A_0A_2, i.e., the point $(A_0A_2)^{\perp(\sigma)}$. Show that if $A_0A_1A_2$ is taken as triangle of reference and E as unit point then the equation of Q is $X_0X_2 - X_1{}^2 = 0$.

9. A line L is a *tangent* to the conic $Q(\sigma)$ if σ_L is degenerate. Show that L is a tangent to a non-degenerate conic Q if, and only if, $L = A^\perp$ for some point A on Q. Find the tangent to the conic $X_0X_2 - X_1{}^2 = 0$ at the point $(\theta^2, \theta\varphi, \varphi^2)$.

10. (See exercise 9.) If Q is a non-empty non-degenerate conic and L a tangent to Q at the point A, we regard L as a set of points and define a mapping φ_L of Q into L as follows: $A\varphi_L = A$; and if P is any point of Q distinct from A then $P\varphi_L$ is the intersection of L with the tangent at P. Show that φ_L is a one–one mapping of Q onto L. If M is any other tangent to Q show that the mapping $\varphi_L{}^{-1}\varphi_M$ yields a projectivity of L onto M.

This example shows that a non-empty non-degenerate conic can be regarded as an abstract one-dimensional projective space by means of the link φ_L and that this space is essentially independent of L.

11. An element M of $\mathscr{P}(V)$ is *tangent* to a quadric $Q(\sigma)$ if σ_M is degenerate. A line L is a *generator* of $Q(\sigma)$ if $\sigma_L = 0$.

Let $Q(\sigma)$ be a non-degenerate quadric in $\mathscr{P}(\mathbf{C}^4)$. Show that a line L cuts $Q(\sigma)$ in a pair of distinct points if, and only if, L and $L^\perp$ are skew; that L is a tangent line, but not a generator if, and only if, $L \cap L^\perp$ is a point; and that L is a generator if, and only if, $L = L^\perp$. If M is a tangent plane, show that the intersection of $Q(\sigma)$ and $\mathscr{P}(M)$ is a pair of generators which intersect in the point $M^\perp$.

12. Show that a non-degenerate quadric $Q(\sigma)$ in $\mathscr{P}(\mathbf{R}^4)$ has no points if the signature of σ is ± 4; has points but no generating lines if the signature of σ is ± 2; has generating lines if the signature of σ is 0.

13. Show how to choose a frame of reference for a projective geometry of dimension 3 so that three given skew lines have equations

$$X_0 = X_2 = 0; \quad X_1 = X_3 = 0; \quad X_0 - X_1 = X_2 - X_3 = 0.$$

Show that the lines meeting these three lines are generators of the quadric $X_0X_3 - X_1X_2 = 0$ and cut the three lines in projective ranges of points.

14. If a non-degenerate quadric Q in a projective geometry of dimension 3 contains one generator L show that every plane through L cuts Q in L and another generator. Show that a coordinate system can be chosen with respect to which Q has equation $X_0X_3 - X_1X_2 = 0$ and L has equations $X_0 = X_1 = 0$.

15. Let $\perp (\sigma)$ be a polarity of an n-dimensional projective geometry P. A simplex of reference $(A_0, \ldots, A_n)$ is called *self-polar* if, for each i, the polar hyperplane $A_i{}^\perp$ of A_i is the join of all the other vertices A_j, $j \neq i$.

Interpret the proof of Theorem 3 (in the case where σ is a non-degenerate symmetric bilinear form and F is not of characteristic 2) as the construction of a self-polar simplex of reference.

16. Let $\mathscr{P}(V)$ be a projective geometry of dimension n and let $Q(\sigma)$ be a quadric in $\mathscr{P}(V)$ containing at least one point outside $V^\perp$. If P is a point not on Q show that there exist n lines through P each of which intersects Q and whose join is V. Give a construction for the polar hyperplane $P^\perp$ of P. (Note that the polar hyperplane of P may not have any intersections with Q.)

Show also how to construct the polar of an arbitrary point of Q.

17. Let σ be a degenerate symmetric bilinear form on V and let M be a subspace of V such that $V = M \oplus V^\perp$. If the quadric $Q(\sigma)$ contains at least one point outside $V^\perp$ show that $Q(\sigma_M)$ is a non-empty non-degenerate quadric and that $Q(\sigma)$ consists of all the points on the lines joining the points of $V^\perp$ to the points of $Q(\sigma_M)$. If P is a point in M show that

$$P^{\perp(\sigma)} = P^{\perp(\sigma_M)} \oplus V^\perp.$$

5.7　Affine Quadrics

All vector spaces in this section are over fields which are not of characteristic 2.

If σ is a symmetric bilinear form on V then the projective quadric $Q(\sigma)$ consists of all points $[a]$ of $\mathscr{P}(V)$ for which $\sigma(a, a) = 0$. The set of all points $\{a\}$ of $\mathscr{A}(V)$ which satisfy $\sigma(a, a) = 0$ is called the *representative cone* of Q and written $C(Q)$. It is clear that if Q is not empty then $C(Q)$ is "generated" by lines of $\mathscr{A}(V)$ through the point O. For example, if Q is a (non-empty) non-degenerate conic in $\mathscr{P}(V)$ then $C(Q)$ is a cone with vertex O in the classical sense. If Q is a pair of (projective) lines then $C(Q)$ is a pair of (affine) planes through O.

The classical definition of ellipses, hyperbolas and parabolas as conic sections suggests that we consider the points in a coset $c+H$ ($c \notin H$) which lie on $C(Q)$. Now

$$\sigma(c+h, c+h) = \sigma(h, h) + 2\sigma(c, h) + \sigma(c, c)$$

may be written in the form

$$q(h) + l(h) + z$$

where q is a quadratic form on H, l is a linear form on H and z is a scalar. Thus the "section" of $C(Q)$ by the coset $c+H$ is the set of all points $\{c+h\}$ where

$$q(h) + l(h) + z = 0 \quad (h \in H).$$

DEFINITION. Let M be a subspace of V. If q is a quadratic form on M, l a linear form on M and z a scalar, then the set of all points $\{a\}$ in $\mathscr{A}(M)$ for which

$$q(a) + l(a) + z = 0$$

is an *affine quadric* in $\mathscr{A}(M)$.

We note immediately that the image of an affine quadric in $\mathscr{A}(M)$ under a translation in $\mathscr{A}(M)$ is again an affine quadric: for if m is a fixed element of M, then

$$q(m+a) - q(m) - q(a) = l_1(a)$$

defines a linear form l_1 on M and so

$$q(a) + l(a) + z = q(m+a) - q(m) - l_1(a) + l(a) + z$$
$$= q(m+a) + l'(m+a) + z',$$

where $l' = l - l_1$ and $z' = q(m) - l(m) + z$ since $l_1(m) = 2q(m)$.

We may now define unambiguously an *affine quadric in* $\mathscr{A}(c+M)$ to be the image of a quadric in $\mathscr{A}(M)$ under the translation by c.

Thus the representative cone $C(Q)$ and all its cross-sections by the elements of $\mathscr{A}(V)$ are examples of affine quadrics.

If φ is a one–one mapping of an affine geometry $\mathscr{A}(S)$ onto a set A then we regard A as an affine geometry by means of the link φ (as on p. 34). An affine quadric in $\mathscr{A}(S)$ is carried over by φ onto a set of points in A which we call an *affine quadric in* A. In particular if A is the affine geometry in $\mathscr{P}(V)$ determined by the hyperplane H (linked to $\mathscr{A}(c+H)$ by the mapping φ of the Embedding Theorem (p. 32)) and if Q is a projective quadric in $\mathscr{P}(V)$, then the set of points in A obtained from Q by omitting its points at infinity (if any) is an affine quadric Q'

in A. We note at once that there may be several different projective quadrics Q which yield the same affine quadric Q' in this way. In other words there may not be a unique way of "completing" an affine quadric Q' in A to a projective quadric Q in $\mathscr{P}(V)$. We shall return to this question in Proposition 13.

Our definitions ensure that any translation carries an affine quadric onto an affine quadric. In fact we have the counterpart of Proposition 10 of the previous section, the proof being almost the same.

PROPOSITION 11. *The image of an affine quadric under an affine isomorphism is an affine quadric.*

(The question of the rank of an affine quadric will be considered later.)

If α is an affine coordinate system for any affine geometry A of dimension n over a field F then an affine quadric in A is the set of all points whose coordinates with respect to α satisfy an equation

$$g(X_1, \ldots, X_n) = 0$$

where g is a polynomial in $F[X_1, \ldots, X_n]$ of degree ≤ 2. In this way the discussion of affine quadrics in A may be reduced to the discussion of affine quadrics in $\mathscr{A}(F^n)$, or in any other convenient affine geometry of dimension n over F.

If H is a hyperplane in a vector space V of dimension $n+1$ over F, then the affine geometry A determined by H in $\mathscr{P}(V)$ is a suitable model for all affine geometries of dimension n over F. Let $(a_1, \ldots, a_n)$ be an ordered basis of H and extend it to an ordered basis $(a_0, \ldots, a_n)$ of V. This induces a coordinate system of $\mathscr{A}(V)$ in which the hyperplane H has equation $X_0 = 0$ and the coset $a_0 + H$ has equation $X_0 = 1$. We now obtain a coordinate system of $\mathscr{A}(a_0 + H)$ and also one of A by assigning to the point $\{a_0 + x_1 a_1 + \cdots + x_n a_n\}$ of $\mathscr{A}(a_0 + H)$ and the corresponding point $[a_0 + x_1 a_1 + \cdots + x_n a_n]$ of A the coordinate row $(x_1, \ldots, x_n)$.

If $f(X_0, \ldots, X_n)$ is a quadratic polynomial with coefficients in F (i.e., a homogeneous polynomial of degree 2 or the zero polynomial), then the equation $f(X_0, \ldots, X_n) = 0$ defines a projective quadric Q in $\mathscr{P}(V)$ that determines the affine quadric Q' in A with equation

$$f(1, X_1, \ldots, X_n) = 0.$$

(Equivalently, $f(X_0, \ldots, X_n) = 0$ defines the representative cone $C(Q)$ in $\mathscr{A}(V)$ whose section by $a_0 + H$ is the affine quadric with equation $f(1, X_1, \ldots, X_n) = 0$.)

If $g(X_1, \ldots, X_n)$ is an arbitrary polynomial in $F[X_1, \ldots, X_n]$ of degree ≤ 2 we may define a quadratic polynomial $f(X_0, \ldots, X_n)$ by the rule

$$f(X_0, \ldots, X_n) = X_0{}^2 \, g(X_1/X_0, \ldots, X_n/X_0)$$

from which follows $f(1, X_1, \ldots, X_n) = g(X_1, \ldots, X_n)$. Thus there is always at least one quadric Q in $\mathscr{P}(V)$ that is a completion of a given quadric Q' in A.

In view of these observations we may prove

PROPOSITION 12. *Given an affine quadric in an n-dimensional affine geometry A, then a coordinate system for A can be chosen with respect to which the quadric has an equation of one of the following two types:*

$$d_1 X_1{}^2 + \cdots + d_n X_n{}^2 + d_0 = 0, \tag{1}$$

$$X_1 + d_2 X_2{}^2 + \cdots + d_n X_n{}^2 = 0 \tag{2}$$

(where some, or all, of the d_i's may be zero).

PROOF. Let A be the affine geometry determined by the hyperplane H in $\mathscr{P}(V)$ and let $Q(\sigma)$ be a quadric in $\mathscr{P}(V)$ that determines the given affine quadric Q' in A.

(i) If $H^\perp \not\subset H$ we choose any vector a_0 in $H^\perp$ but not in H and any ordered basis $(a_1, \ldots, a_n)$ of mutually orthogonal vectors for H (using Theorem 3 on p. 103). The equation of Q is then of the form

$$d_0 X_0 + \cdots + d_n X_n{}^2 = 0$$

which gives equation (1) when we substitute $X_0 = 1$.

(ii) If $H^\perp \subset H$ then certainly $V^\perp \subset H$ and so Proposition 4 (p. 93) gives $\dim H + \dim H^\perp = \dim V + \dim V^\perp$, i.e., $\dim V^\perp = \dim H^\perp - 1$. Let A be any point in $H^\perp$ but not in $V^\perp$. Any line L through A not in $H = A^\perp$ intersects Q in just one other point, say B. (Cf. the discussion of quadrics on a line, p. 112. See also exercise 5 of § 5.6.) The restriction of σ to L is non-degenerate and so, by Proposition 5 (p. 96), $V = L \oplus L^\perp$. We choose arbitrary homogeneous vectors a_0, a_1 for B, A, respectively, and express $L^\perp$ as direct sum $[a_2] \oplus \cdots \oplus [a_n]$ of mutually orthogonal 1-dimensional subspaces (again using Theorem 3). The equation of Q in the resulting coordinate system for $\mathscr{P}(V)$ is of the form

$$X_0 X_1 + d_2 X_2{}^2 + \cdots + d_n X_n{}^2 = 0$$

which yields (2) if we substitute $X_0 = 1$.

Many of the properties of an affine quadric Q' in A are most easily understood in terms of a projective quadric Q completing Q' in $\mathscr{P}(V)$. It is essential therefore to find out just when Q is uniquely determined by Q'.

PROPOSITION 13. *Let A be the affine geometry determined by the hyperplane H in $\mathscr{P}(V)$. If Q' is an affine quadric in A which is not contained in any hyperplane of A, then there is one and only one quadric Q in $\mathscr{P}(V)$ whose intersection with A is Q'.*

PROOF. We have seen above that there is always at least one quadric Q completing Q'. What has to be done now is to show that the points of Q in H are determined by Q'. We do this by proving that a point P of H does *not* lie on Q if, and only if, there is a line through P meeting Q' in exactly two points.

If P lies on Q then any line L through P which meets Q' in two points contains three points of Q and so L lies entirely on Q. The ground field F contains at least three elements and therefore L contains at least four points, three of which lie on Q'. Conversely, if P does not lie on Q then $P^{\perp(\sigma)}$ cannot be the whole of V and so must be a hyperplane. By our hypothesis on Q', we may choose a point, say M, on Q' but not in this hyperplane. Then PM meets Q in just one point other than M (cf. the discussion of quadrics on a line, p. 112). This point must clearly lie on Q'.

For any quadric Q' not contained in a hyperplane we define the *rank* of Q' to be the rank of the corresponding projective quadric Q and say that Q' is *degenerate* if, and only if, Q is degenerate.

The most obvious case not covered by Proposition 13 is when Q' is the set of *all* points of a hyperplane. Suppose this hyperplane has equation $X_1 = 0$. We obviously cannot distinguish between the sets of points defined by the equations $X_1 = 0$ and $X_1^2 = 0$. A projective quadric completing Q' must then be either $X_0 X_1 = 0$ or $X_1^2 = 0$.

The determination of all affine quadrics in A which are contained in a hyperplane is not difficult (see exercise 7, below). We mention that there are only two cases where a non-empty *non-degenerate* quadric Q determines a quadric Q' which is contained in a hyperplane:

(1) Q is a pair of points on a projective line, one of which is the point (hyperplane) at infinity H.

(2) $\mathscr{P}(V)$ is a projective plane over the field $\mathbf{F}_3$ and Q is a non-degenerate conic cutting the line at infinity in two points, so that Q' consists of just two points. (There are exactly three conics Q determining Q' in this case. Cf. exercise 8, below.)

We may obviously exclude these two cases by assuming that our geometry has dimension $n \geq 2$ and that Q' contains at least three points. This would allow us for example to refer to the point $H^\perp$ (where H is the hyperplane at infinity and $\perp$ is the polarity determined by Q, cf. exercise 16 of § 5.6) as the *center* of Q' and to call Q' a *paraboloid* if this center lies in H (i.e., lies on Q). If we consider Proposition 12 in the non-degenerate case (i.e., the d_i's are all non-zero), then equation (1) defines a quadric with center at the origin $(0, 0, \ldots, 0)$ and equation (2) defines a paraboloid through the origin whose center is the point at infinity on the X_1-axis (i.e., the line joining the origin and $(1, 0, \ldots, 0)$).

Exercises

1. If Q' is the affine conic in A determined by a non-degenerate projective conic Q in $\mathscr{P}(\mathbf{R}^3)$, then Q' is called an *ellipse*, a *parabola* or a *hyperbola* according as the intersection of Q with the line at infinity is empty, consists of one point or consists of two points.   (The empty conic is thus an ellipse.)   Show that an affine conic with equation

$$a_{11}X_1^2 + 2a_{12}X_1X_2 + a_{22}X_2^2 + 2a_{01}X_1 + 2a_{02}X_2 + a_{00} = 0,$$

where the 3×3 matrix (a_{ij}) is invertible, is an ellipse, parabola or hyperbola according as $a_{11}a_{22} - a_{12}^2$ is positive, zero or negative, respectively.

2. Show that the affine polarity of $\mathscr{A}(\mathbf{R}^2)$ (see exercise 5 of § 5.5) which sends the point $(x_1, x_2) \neq (0, 0)$ into the line $x_1X_1 + x_2X_2 + 1 = 0$ has empty conic. (This polarity is the product of a reflexion in the origin followed by the polarity with conic $X_1^2 + X_2^2 = 1$.)

3. If $Q(\sigma)$ is a projective quadric in $\mathscr{P}(V)$ contained in a hyperplane of $\mathscr{P}(V)$, show that $Q(\sigma)$ is precisely the set of all points in $V^{\perp(\sigma)}$.   (Hint: consider the lines which meet $Q(\sigma)$ but do not lie in H.)   (Note that this gives all the projective quadrics of $\mathscr{P}(V)$ that determine the empty quadric in A.)

4. What are the possible ranks of the quadrics described in exercise 3 when (i) $F = \mathbf{R}$, (ii) $F = \mathbf{C}$, (iii) $F = \mathbf{F}_3$?

5. If Q is a non-empty *non-degenerate* quadric in $\mathscr{P}(V)$ contained in a pair of hyperplanes show that either (i) pdim $V = 1$ and Q consists of two points or (ii) the ground field is $\mathbf{F}_3$, pdim $V = 2$ and Q consists of four points.

6. Use exercise 5 above and exercise 17 of § 5.6 to find all the degenerate projective quadrics that are contained in a pair of hyperplanes.

7. Use exercises 3, 5, 6 to find all the affine quadrics in A that lie in a hyperplane of A.

8. If $F = \mathbf{F}_3$ and Q' is an affine conic in A consisting of two points, show that Q' has three "completions" to $\mathscr{P}(V)$ all of which are non-degenerate.

5.8   Sesquilinear Forms

We have seen that to every non-degenerate bilinear form on V there are associated two correlations $\perp(\sigma)$, $\top(\sigma)$ of $\mathscr{P}(V)$.   On the other hand, we know from § 4.7 that a correlation is but a special case of an anti-automorphism.   We shall now show that there exists a generalization of bilinear form that gives rise to anti-automorphisms in the same way that bilinear forms yield correlations.

DEFINITION.   Let V, V' be vector spaces over fields F, F', respectively, and ζ a mapping of F onto F'.   Then a mapping f of V into V' is *semi-linear* with respect to ζ if

L.1. $(a+b)f = af + bf,$

L(ζ).2. $(xa)f = (x\zeta)(af),$

for all a, b in V and all x in F.

If $Vf = 0$ the mapping ζ plays no role. But if $Vf \neq 0$ we may prove exactly as before (p. 55) that

$$(x+y)\zeta = x\zeta + y\zeta,$$

$$(xy)\zeta = (x\zeta)(y\zeta)$$

for all x, y in F. It follows that ζ is an isomorphism of F onto F' (cf. exercise 1 of § 3.6 (p. 62)).

Let f be a semi-linear mapping of V into V^* with respect to an automorphism ζ of F. We define a mapping σ by the rule

$$\sigma(a, b) = b(af)$$

for all a, b in V. Since $af \in V^*$, σ is linear in the second variable, but is semi-linear with respect to ζ in the first variable. Explicitly,

B(ζ).1. $\sigma(xa + yb, c) = (x\zeta) \, \sigma(a, \cdot c) + (y\zeta) \, \sigma(b, c),$

B.2. $\sigma(a, xb + yc) = x \, \sigma(a, b) + y \, \sigma(a, c),$

for all a, b, c in V and all x, y in F. Such a mapping σ is called a *left sesquilinear form* with respect to ζ.

We may interchange the two variables and define a mapping τ by the equation

$$\tau(a, b) = a(bf).$$

Then τ is called a *right* sesquilinear form with respect to ζ. The theory of these is exactly like that of the left sesquilinear forms. We shall consider only the latter and shall usually omit the qualifying adjective "left". Clearly, a sesquilinear form with respect to the identity mapping is the same as a bilinear form.

It is worth noting that if σ is a left sesquilinear form with respect to ζ then $\sigma\zeta^{-1}$ is a right sesquilinear form with respect to ζ^{-1}. (As usual, $\sigma\zeta^{-1}$ denotes a mapping product: thus $(\sigma\zeta^{-1})(a, b) = \sigma(a, b)\zeta^{-1}$.) More generally, if σ is semi-linear with respect to ξ in the first variable and semi-linear with respect to η in the second variable, then $\sigma\eta^{-1}$ is left sesquilinear with respect to $\xi\eta^{-1}$ and $\sigma\xi^{-1}$ is right sesquilinear with respect to $\eta\xi^{-1}$. As far as "orthogonality" properties are concerned these forms would give no further generality.

Let σ be a (left) sesquilinear form with respect to ζ. Then we define g just as before by the equation

$$\sigma(a, b) = b(ag)$$

and check that g is a semi-linear mapping with respect to ζ of V into V^*. In order that the companion mapping $\bar{\sigma}$ should take V into V^* we must use the form $\sigma\zeta^{-1}$, which is right sesquilinear with respect to ζ^{-1}, and set

$$\sigma(a, b)\zeta^{-1} = a(b\tilde{\sigma})$$

for all a, b in V. Then $\tilde{\sigma}$ is semi-linear with respect to ζ^{-1}.

The notion of orthogonality with respect to σ and the mappings $\perp(\sigma)$, $\top(\sigma)$ are defined exactly as in the bilinear case. *All the results of § 5.2 remain true for sesquilinear forms.* Moreover the proofs are precisely as before. In establishing the sesquilinear version of Proposition 4 one needs the extension of Theorem 1, Chapter IV (p. 66), to semi-linear mappings, but this is quite obvious. We then *define* the rank of σ to be $\dim V - \dim V^{\perp}$ and say that σ is non-degenerate if $V^{\perp} = 0$. Theorem 1 now gives the relationship between the anti-automorphisms $\perp(\sigma)$ and $\top(\sigma)$ in the non-degenerate case. If we assume that $\mathrm{pdim}\, V \geq 2$ then any anti-automorphism of $\mathscr{P}(V)$ is of the form $\mathscr{P}(f) \circ$, where f is a semi-linear isomorphism (p. 85) and so can be expressed as $\perp(\sigma)$ or $\top(\sigma)$ in view of the extended Lemma 1.

The question of orthosymmetry for sesquilinear forms is more subtle. The condition $\perp(\sigma) = \top(\sigma)$ is equivalent to $\mathscr{P}(\sigma) = \mathscr{P}(\tilde{\sigma})$ as before. We now need to assume that $\mathrm{rank}\,\sigma \geq 2$. Then arguing as in Lemma 4 (p. 97), we deduce $\sigma = z\tilde{\sigma}$ for some $z \neq 0$. (Note that we now invoke Proposition 1 of Chapter III (p. 45) for semi-linear isomorphisms: the validity of this has already been remarked upon (p. 57).) Two things now follow: first of all, $\zeta = \zeta^{-1}$, i.e., $\zeta^2 = 1$, the identity mapping on F; and secondly

$$\sigma(a, b) = z(\sigma(b, a)\zeta) = z(z\zeta)\sigma(a, b)$$

shows that $z(z\zeta) = 1$, the identity element of F.

If $\sigma(a, a) = 0$ for all a then we shall see below (Lemma 6) that σ is actually alternating. Suppose now, however, that $\sigma(c, c) = w \neq 0$ for some c. Then $w = z(w\zeta)$ and so if we write

$$\tau(a, b) = (w\zeta)\,\sigma(a, b)$$

for all a, b, we have

$$\tau(a, b) = \tau(b, a)\zeta$$

for all a, b.

DEFINITION. A sesquilinear form σ with respect to ζ is called *hermitian* if

$$\sigma(a, b) = \sigma(b, a)\zeta$$

for all a, b. (If $\sigma \neq 0$ this implies that $\zeta^2 = 1$.)

LEMMA 6. *If the sesquilinear form σ satisfies $\sigma(a, a) = 0$ for all a in V, then σ is bilinear (and hence alternating).*

PROOF. For any a, b in V,

$$0 = \sigma(a+b, a+b) = \sigma(a, b) + \sigma(b, a)$$

and hence, for all x in F,

$$x\sigma(a, b) = \sigma(a, xb) = -\sigma(xb, a) = -(x\zeta)\sigma(b, a) = (x\zeta)\sigma(a, b).$$

If $\sigma \neq 0$ we conclude $x = x\zeta$ and so σ is bilinear.

To sum up, we have proved

PROPOSITION 14. *An orthosymmetric sesquilinear form of rank ≥ 2 is either alternating or is a scalar multiple of a hermitian form.*

The lemma shows that if we wish to extend the structure theorems of § 5.4 to sesquilinear forms we need only consider forms that are not alternating. *Theorem 3 is true for (non-alternating) hermitian forms and Theorem 4 is true for arbitrary (but non-alternating) sesquilinear forms.* The proofs of both theorems are unchanged up to the point where σ is assumed to be non-zero and alternating on $[a]^{\perp}$. But this implies that the automorphism of the restriction of σ to $[a]^{\perp}$, and so also of σ itself, is the identity mapping. Thus σ is bilinear and we are genuinely back in the situations of the earlier theorems.

We have held back till now the obvious remark that for the fields $\mathbf{F}_p$ (p any prime), $\mathbf{Q}$ and $\mathbf{R}$ all sesquilinear forms are necessarily bilinear (because the only automorphism of each of these fields is the identity mapping). The field $\mathbf{C}$ has one non-identity automorphism of outstanding importance: this is complex conjugation. It is the only non-identity automorphism of $\mathbf{C}$ that leaves all the real numbers fixed (cf. exercise 1, § 3.5, p. 58). When hermitian (but not bilinear) forms on complex vector spaces are under consideration it is usual to make the convention that the relevant automorphism is complex conjugation (denoted by $x \rightarrow \bar{x}$).

If σ is a hermitian form on the complex vector space W (with respect to complex conjugation) then by the generalized version of Theorem 3 there exists an ordered basis $(a_1, \ldots, a_n)$ such that $\sigma(a_i, a_j) = 0$ for $i \neq j$. Under what circumstances can we choose the basis so that in addition $\sigma(a_i, a_i) = 1$ for all i? We look for scalars $x_1, \ldots, x_n$ such that

$$\sigma(x_i a_i, x_i a_i) = \bar{x}_i x_i \sigma(a_i, a_i) = 1.$$

These can be found if, and only if, each $\sigma(a_i, a_i)$ is real and positive. But this implies that $\sigma(a, a)$ is real and positive for *all* non-zero vectors a in W.

DEFINITION. A hermitian form σ on a complex vector space W with respect to complex conjugation is called *positive definite* if $\sigma(a, a) > 0$ for all non-zero vectors a in W.

Let V be a real vector space and $V_{(C)}$ its complexification. If σ is a bilinear form on V we may extend σ to a mapping $\sigma_{(C)}$ by the rule

$$\sigma_{(C)}(a+ib, c+id) = \sigma(a, c) + \sigma(b, d) + i(\sigma(a, d) - \sigma(b, c))$$

for all $a + ib$, $c + id$ in $V_{(C)}$. The reader may check that $\sigma_{(C)}$ is a sesquilinear form on $V_{(C)}$ (with respect to complex conjugation). We shall call $\sigma_{(C)}$ the *complexification* of σ.

The above discussion shows that $\sigma_{(C)}$ is positive definite if, and only if, the real form σ is positive definite.

The reader should compare the extension $\sigma_{(C)}$ with the extension σ' used in § 5.6 (p. 113). Of course σ' is bilinear, but the property of positive definiteness is lost in the passage from σ to σ'.

EXERCISES

Notation. If $A = (a_{ij})$ is a matrix with elements a_{ij} in F and if ζ is an automorphism of F, then $A\zeta$ denotes the matrix $(a_{ij}\zeta)$.

1. Let σ be a sesquilinear form with respect to ζ on V and $(a_1, \ldots, a_n)$ an ordered basis of V. Define the matrix S of σ with respect to $(a_1, \ldots, a_n)$ as for bilinear forms. Prove that if

$$x = (x_1, \ldots, x_n), \quad y = (y_1, \ldots, y_n)$$

are the coordinate rows of a, b with respect to $(a_1, \ldots, a_n)$, then

$$\sigma(a, b) = (x\zeta)Sy^t.$$

2. Show that S is the matrix of a hermitian form if, and only if,

$$S = (S\zeta)^t$$

for some automorphism ζ satisfying $\zeta^2 = 1$. (We call such a matrix S *hermitian*.)

3. If σ is an alternating bilinear form on the real vector space V show that $\sigma_{(C)}$ is i times a hermitian form on $V_{(C)}$ $(i^2 = -1)$.

4. Let σ be a sesquilinear form on V with respect to ζ and f a linear automorphism of V. If (a_i) is an ordered basis of V and

$$(f; (a_i)) = P, \quad (\sigma; (a_i)) = S,$$

prove that f satisfies

$$\sigma(af, bf) = \sigma(a, b)$$

if, and only if,

$$S = (P\zeta)SP^t.$$

(We call such a mapping f an automorphism of the sesquilinear space (V, σ).)

CHAPTER VI

Euclidean Geometry

6.1 Distances and Euclidean Geometries

DEFINITION. If V is a real vector space and σ is a positive definite bilinear form on V, then (V, σ) is called a *euclidean space*.

For any vectors a, b in a euclidean space (V, σ) we define

$$\|a\| = +\sqrt{\sigma(a, a)}, \qquad d(a, b) = \|a-b\|$$

and call $\|a\|$ the *length* (or *norm*) of a, $d(a, b)$ the *distance* from a to b.

Then d has the following basic properties: For all a, b, c in V,

D.1. $d(a, b) = d(b, a)$;

D.2. $d(a, b) \geq 0$ and $d(a, b) = 0$ if, and only if, $a = b$;

D.3. $d(a, c) \leq d(a, b) + d(b, c)$ *(the triangle inequality)*;

D.4. $d(a+c, b+c) = d(a, b)$;

D.5. for any real x, $d(xa, xb) = |x|\, d(a, b)$;

D.6. $d(a+b, 0)^2 + d(a-b, 0)^2 = 2(d(a, 0)^2 + d(b, 0)^2)$

$$(the\ parallelogram\ law).$$

D.1, D.2, D.4 and D.5 are all obvious; D.3 is proved below and D.6 follows from an easy calculation which we leave to the reader. The reason for calling D.6 the parallelogram law is plain: the affine points a, 0, b, $a+b$ are the vertices of a parallelogram and D.6 asserts that the sum of the squares of the two diagonals equals the sum of the squares of all four sides.

DEFINITION. Let V be a real vector space and d a mapping of the ordered pairs of vectors of V into $\mathbf{R}$. If d satisfies D.1–D.6, then d is called a *distance* on V.

We remark in passing that a mapping d which satisfies D1, D.2, D.3 is called a *metric* on V and (V, d) is a *metric space*. If the metric d also satisfies D.4 and D.5 then (V, d) is called a *normed space* and the mapping $a \to \|a\|$ where $\|a\| = d(a, 0)$ is its *norm*. (Cf. exercises 1, 2 below.)

LEMMA 1. (SCHWARZ'S INEQUALITY.) *For all a, b in the euclidean space (V, σ),*

$$|\sigma(a, b)| \leq \|a\| \, \|b\|.$$

Equality holds if, and only if, a and b are linearly dependent.

PROOF. The result is clear when $a = 0$. So assume $\|a\| \neq 0$. We have $\sigma(xa - b, xa - b) = \|a\|^2 x^2 - 2\sigma(a, b)x + \|b\|^2$, and since the left-hand side is never negative, the substitution $x = \sigma(a, b)/\|a\|^2$ gives $\sigma(a, b)^2 \leq \|a\|^2 \|b\|^2$. Further, equality holds if, and only if, $b = xa$.

PROOF OF D.3. For any a, b in V,

$$\begin{aligned}
\|a + b\|^2 &= \sigma(a + b, a + b) \\
&= \|a\|^2 + \|b\|^2 + 2\sigma(a, b) \\
&\leq \|a\|^2 + \|b\|^2 + 2\|a\| \, \|b\|, \quad \text{by Lemma 1,} \\
&= (\|a\| + \|b\|)^2.
\end{aligned}$$

Thus $\|a + b\| \leq \|a\| + \|b\|$, which is equivalent to D.3.

In addition to the distance between two vectors in a euclidean space we may also introduce the angle between them.

Let $\cos \theta$ be defined for all real θ as the sum of the series

$$1 - \frac{\theta^2}{2!} + \frac{\theta^4}{4!} - \cdots$$

and recall from real analysis the fact that cos is a strictly decreasing mapping of the closed interval $[0, \pi]$ onto the closed interval $[-1, 1]$. Thus we may find a unique θ in $[0, \pi]$ satisfying the equation

$$\cos \theta = \sigma(a, b)/\|a\| \, \|b\|$$

for any given non-zero vectors a, b in (V, σ) (because the right-hand side is in $[-1, 1]$, by Lemma 1). The real number θ is called the *angle between a and b*. Note that the special case $\theta = \pi/2$ explains the terminology of § 5.2!

Every positive definite bilinear form determines a distance. Conversely, suppose a distance d is given on the real vector space V. If we set $\sigma(a, a) = d(a, 0)^2$ and

$$2\sigma(a, b) = \sigma(a + b, a + b) - \sigma(a, a) - \sigma(b, b),$$

then it may be shown (cf. exercises 4, 5 below) that σ is a positive definite bilinear form on V so that (V, σ) is a euclidean space. We shall also write this as (V, d). It is therefore immaterial whether we use a positive definite bilinear form or a distance to define euclidean space.

If σ is a positive definite bilinear form on V, then we know from Proposition 9 of Chapter V (p. 106) that V has an ordered basis $(a_1, \ldots, a_n)$ such that the matrix of σ with respect to $(a_1, \ldots, a_n)$ is I_n, the identity matrix. This basis clearly consists of mutually orthogonal vectors of length 1. We call such a basis a *cartesian basis* of (V, σ). If d is the distance determined by σ and $a = \sum x_i a_i$, $b = \sum y_i a_i$, then

$$d(a, b) = +\sqrt{(x_1 - y_1)^2 + \cdots + (x_n - y_n)^2}.$$

If the vectors a, b are non-zero then the angle θ between them is given by

$$\cos \theta = \frac{x_1 y_1 + \cdots + x_n y_n}{+\sqrt{(x_1{}^2 + \cdots + x_n{}^2)(y_1{}^2 + \cdots + y_n{}^2)}}.$$

Let (V', σ') be a second euclidean space, also of dimension n, and d' the corresponding distance. If f is a linear isomorphism of V onto V', then it is easy to verify that the condition

$$\sigma(a, b) = \sigma'(af, bf)$$

for all a, b in V, is equivalent to

$$d(a, b) = d'(af, bf)$$

for all a, b in V. Hence f is an isomorphism of (V, σ) onto (V', σ') if, and only if, f preserves distances. Moreover, f has this property if, and only if, f maps any cartesian basis of (V, σ) onto a cartesian basis of (V', σ').

Consider now a coset $S = a + M$ in a real vector space V and suppose d is a distance on M. We define a mapping, also denoted by d, of the ordered pairs of vectors of S into $\mathbf{R}$ by

$$d(a + m_1, a + m_2) = d(m_1, m_2).$$

This definition is independent of a because distance is invariant under translations (D.4). We shall call this new mapping a *distance on the coset S*. It is clear that d yields a mapping of the ordered pairs of points of $\mathscr{A}(S)$ into $\mathbf{R}$, namely

$$(\{u\}, \{v\}) \to d(u, v).$$

We also denote this mapping by d and call $d(\{u\}, \{v\})$ the distance between the points $\{u\}$ and $\{v\}$.

Observe that a bilinear form on M could not have been extended unambiguously in this way. This is the reason for preferring the notion of distance to that of bilinear form in geometrical contexts. Note also that if the distance d on M coincides with the restriction to M of a distance d' on V, then d' and d also coincide on S.

DEFINITION. A real affine geometry $\mathscr{A}(S)$ together with a distance d on $\mathscr{A}(S)$ is called a *euclidean geometry* and will be denoted by $\mathscr{A}(S, d)$.

If T is a coset contained in S, the restriction of d to $\mathscr{A}(T)$ is a distance on $\mathscr{A}(T)$ and $\mathscr{A}(T, d)$ is said to be a *subgeometry* of $\mathscr{A}(S, d)$.

We may introduce *angles* in a euclidean geometry as follows.   Given a point A in the euclidean geometry $\mathscr{A}(S, d)$ and two further points B, C, both distinct from A, we define the angle $\widehat{BAC}$ to be the angle between the vectors $b - a$ and $c - a$ where a, b, c are the vectors for A, B, C, respectively.

Now let L, L' be two lines in $\mathscr{A}(S, d)$.   Through any point A on L we take the line L'' parallel to L' and choose points B, C on L, L'', respectively, both distinct from A.   As B and C vary, so $\widehat{BAC}$ takes just two values in $[0, \pi]$ and the sum of the two values is π.   We shall call the value in $[0, \pi/2]$ the angle between L and L'.   Thus the cosine of the angle between L and L' is given by $|\sigma(u, u')|/\|u\| \, \|u'\|$, for any $u \neq 0$ in the subspace belonging to L and any $u' \neq 0$ in the subspace belonging to L'.   (Here σ is the bilinear form giving the distance d.)

An *isomorphism* α of a euclidean geometry $\mathscr{A}(S, d)$ onto a euclidean geometry $\mathscr{A}(S', d')$ is an isomorphism of affine geometries that preserves distance, i.e.,

$$d(u, v) = d'(u\alpha, v\alpha)$$

for all u, v in S.

PROPOSITION 1.   *An isomorphism of euclidean geometries is an affinity whose underlying linear isomorphism is an isomorphism of euclidean spaces.*

PROOF.   Suppose $\dim S \geq 2$ and that M, M' are the subspaces belonging to the cosets S, S', respectively.   By Theorem 5 of Chapter III (p. 56), $\alpha = \mathscr{A}(t_{-a} g \, t_{a\alpha})$, where g is a semi-linear isomorphism of M onto M'.   Since $\mathbf{R}$ has only the identity isomorphism (cf. p. 52), g must be linear.   Hence α is an affinity and g is an isomorphism of euclidean spaces.

It only remains to prove the proposition when $\dim S$ equals 0 or 1. The former case is trivial.   Assume that b is a non-zero element of M and that $a\alpha = a'$, $(a + b)\alpha = a' + b'$.   Then $(a + xb)\alpha = a' + x'b'$ where $x \to x'$ is a one–one mapping of $\mathbf{R}$ onto $\mathbf{R}$.   But α preserves the distances between a, $a + b$ and $a + xb$ so that

$$|x| = |x'| \quad \text{and} \quad |x - 1| = |x' - 1|.$$

This implies that $x' = x$ and so completes the proof.

In order to define coordinate systems for euclidean geometries we must first impose a natural euclidean structure on $\mathbf{R}^n$.   We use the unique bilinear form σ_0 on $\mathbf{R}^n$ for which the standard basis $(e_1, \ldots, e_n)$ is a cartesian basis of $(\mathbf{R}^n, \sigma_0)$. Explicitly, we define

$$\sigma_0((x_1, \ldots, x_n), (y_1, \ldots, y_n)) = x_1 y_1 + \cdots + x_n y_n$$

and denote the corresponding distance by d_0.

DEFINITION. If $\mathscr{A}(S, d)$ is a euclidean geometry of dimension n, then a euclidean isomorphism α of $\mathscr{A}(S, d)$ onto $\mathscr{A}(\mathbf{R}^n, d_0)$ is called a *euclidean coordinate system* for $\mathscr{A}(S, d)$.

The *euclidean frame of reference* corresponding to the coordinate system α is the ordered $(n + 1)$-tuple of points $(P_1, \ldots, P_n, Q)$ where $P_i \alpha = \{e_i\}$, $i = 1, \ldots, n$ and $Q\alpha = \{(0, \ldots, 0)\}$ (cf. p. 43).

If M is the subspace belonging to S we may express the condition for $(P_1, \ldots, P_n, Q)$ to be a frame of reference in a different way. Let $p_1, \ldots, p_n, q$ be vectors for the points $P_1, \ldots, P_n, Q$, respectively. Then $(P_1, \ldots, P_n, Q)$ is a euclidean frame of reference if, and only if, $(p_1 - q, \ldots, p_n - q)$ is a cartesian basis of M (cf. exercise 4 of § 2.2, p. 20).

EXERCISES

1. Define a mapping d of the ordered pairs of vectors of the real vector space V by setting $d(a, a) = 0$ for all a and $d(a, b) = 1$ whenever $a \neq b$. Verify that (V, d) is a metric space but not a normed space.

2. Let $a = (x_1, \ldots, x_n)$, $b = (y_1, \ldots, y_n)$ be elements of $\mathbf{R}^n$ and define

$$d_p(a, b) = (|x_1 - y_1|^p + \cdots + |x_n - y_n|^p)^{1/p}$$

   where $p \geq 1$ is a real number. Show that d_p satisfies all the rules D.1 to D.5, but satisfies D.6 if, and only if, $p = 2$. (Minkowski's inequality

$$\left(\sum |x_i + y_i|^p \right)^{1/p} \leq \left(\sum |x_i|^p \right)^{1/p} + \left(\sum |y_i|^p \right)^{1/p}$$

   (for $p \geq 1$) may be assumed.)

3. Let a, b be vectors in the euclidean space (V, d) and x, y real numbers such that $x + y = 1$. Prove that

$$d(xa + yb, a) : d(b, xa + yb) = |y| : |x|.$$

   (Observe that this is consistent with our remark on p. 29.)

4. Let d be a mapping into $\mathbf{R}$ of the ordered pairs of vectors in the real vector space V. Define a mapping σ by

$$\sigma(a, b) = \tfrac{1}{2}(d(a + b, 0)^2 - d(a, 0)^2 - d(b, 0)^2).$$

   If d satisfies rule D.6 (p. 125), prove that

$$\sigma(a_1 + a_2, b) = \sigma(a_1, b) + \sigma(a_2, b)$$

and deduce that

$$\sigma(ra, b) = r\sigma(a, b)$$

for every rational number r.

(Similar results apply to the second variable. Hence σ is a bilinear mapping of V regarded as a vector space over $\mathbf{Q}$. Note that V is necessarily infinite dimensional over $\mathbf{Q}$ if $V \neq 0$.)

5. Using the same notation as in exercise 4, assume now that d satisfies all the rules D.1–D.6. Prove that, for fixed a, b in V,

$$x \to \sigma(xa, b)$$

is a continuous real function. Deduce from this and exercise 4 that σ is a bilinear form on the real vector space V.

6. Let ABC be a triangle in a euclidean plane with distance d. Prove that

$$2\,d(A, B)\,d(A, C)\cos \widehat{BAC} = d(A, B)^2 + d(A, C)^2 - d(B, C)^2.$$

7. If $d(a, c) = d(a, b) + d(b, c)$ show that a, b, c are collinear in $\mathcal{A}(S, d)$.

If φ is a one–one mapping of the points of $\mathcal{A}(S, d)$ onto the points of $\mathcal{A}(S', d')$ which preserves distance, show that φ induces a euclidean isomorphism of $\mathcal{A}(S, d)$ onto $\mathcal{A}(S', d')$.

6.2   Similarity Euclidean Geometries

We have constantly used the embedding of affine geometry in projective geometry. If our embedded geometry is euclidean, we naturally hope to find an interpretation of the euclidean structure in projective geometrical terms.

Let us recall some facts from § 2.6. Suppose H is a hyperplane in the real projective geometry $\mathscr{P}(V)$ and A is the set of subspaces of V not contained in H. Choose c to be any vector not in H. Then every coset of H can be written uniquely in the form $xc + H$, for x in $\mathbf{R}$. To each x in $\mathbf{R}^*$ (the non-zero real numbers) we associate the mapping $\varphi_x : S \to [S]$ of $\mathcal{A}(xc + H)$ onto A. Then (A, φ_x) is an affine geometry in the generalized sense and all these geometries, for varying non-zero x, have the same basic geometric operation of inclusion. This is equivalent to the fact that, for all x in $\mathbf{R}^*$, $\varphi_1\varphi_x{}^{-1}$ is an affine isomorphism of $\mathcal{A}(c + H)$ onto $\mathcal{A}(xc + H)$. Thus all the geometries (A, φ_x) could be identified and A itself viewed unambiguously as an affine geometry.

Now suppose d is a distance on H. Then each coset $xc + H$ has on it a distance which we shall denote by d_x. We thus obtain a family of euclidean geometries $\mathcal{A}(xc + H, d_x)$ and are naturally interested in knowing whether each $\varphi_1\varphi_x{}^{-1}$ preserves distance.

Choose $a_1 = c + h_1$ and $a_2 = c + h_2$ in $c + H$. Then $a_i \, \varphi_1 \varphi_x^{-1} = xc + xh_i$ for $i = 1, 2$ and hence

$$d_x(a_1 \varphi_1 \varphi_x^{-1}, a_2 \varphi_1 \varphi_x^{-1}) = |x| \, d(h_1, h_2) = |x| \, d_1(a_1, a_2).$$

It follows that $\varphi_1 \varphi_x^{-1}$ is nearly, but not quite, a euclidean isomorphism: it distorts distance by the constant factor $|x|$. Our calculation suggests the following useful generalization of euclidean geometry.

DEFINITION. Two distances d, d' on an affine geometry $\mathscr{A}(S)$ are called *similar* if $d' = rd$ for some positive real number r.

Similarity is an equivalence relation on the set of all distances on $\mathscr{A}(S)$. We shall denote the class containing d by (d).

DEFINITION. Let S be a coset in the real vector space V and D a similarity class of distances on $\mathscr{A}(S)$. The affine geometry $\mathscr{A}(S)$ together with D is called a *similarity euclidean geometry* and will be denoted by $\mathscr{A}(S, D)$.

If T is a coset contained in S, the restriction of all the distances in D to $\mathscr{A}(T)$ is a similarity class of distances on $\mathscr{A}(T)$. We call $\mathscr{A}(T, D)$ a *subgeometry* of $\mathscr{A}(S, D)$.

It is clear that a euclidean geometry $\mathscr{A}(S, d)$ gives rise to a similarity euclidean geometry $\mathscr{A}(S, (d))$.

We now introduce a generalization just as we did for projective and affine geometries. Let φ be a one–one mapping of some $\mathscr{A}(S)$ onto a set A. If d is a distance on $\mathscr{A}(S)$ then φ enables us to define a distance d_φ on A: we simply set

$$d_\varphi(P\varphi, Q\varphi) = d(P, Q)$$

for all points P, Q in $\mathscr{A}(S)$. If D is a similarity class of distances on $\mathscr{A}(S)$ then D_φ denotes the corresponding class on A.

The affine geometry (A, φ) together with a similarity class of distances D_φ is called a *similarity euclidean geometry* and will be denoted by (A, D_φ).

We naturally view $\mathscr{A}(S, D)$ itself as a geometry in this generalized sense by taking φ to be the identity mapping.

An *isomorphism* α of the similarity euclidean geometry $\mathscr{A}(S, D)$ onto the similarity euclidean geometry $\mathscr{A}(S', D')$ is an affine isomorphism such that, for any d in D, the mapping d' defined by

$$d'(u\alpha, v\alpha) = d(u, v)$$

is a distance in D'.

Returning to the embedding situation discussed above, we have a family of similarity euclidean geometries $\mathscr{A}(xc + H, (d_x))$, one for each x

in $\mathbf{R}^*$. Our calculation may be interpreted as showing that each $\varphi_1 \varphi_x{}^{-1}$ is a similarity euclidean isomorphism and so the distance classes $(d_1)_{\varphi_1}$ and $(d_x)_{\varphi_x}$ on A are equal. If D is the distance class of d on H then we also denote by D this unique distance class on A, and we shall refer to (A, D) as the *similarity euclidean geometry in* $\mathscr{P}(V)$ *determined by the hyperplane H and the distance class D.*

We must now ask what it means projectively to be given a distance class D on $\mathscr{A}(H)$. If d and d' are distances arising from positive definite bilinear forms σ and σ', respectively, then it is easy to check that, for any $r > 0$, $d' = rd$ if, and only if, $\sigma' = r^2\sigma$. Hence the bilinear forms associated with D all determine the same polarity of $\mathscr{P}(H)$. The quadric of this polarity is, of course, empty.

DEFINITION. A polarity of a real projective geometry is called *definite* if the corresponding quadric is empty.

PROPOSITION 2. *The real polarity* $\perp(\sigma)$ *is definite if, and only if, the bilinear form σ is either positive definite or negative definite.*

PROOF. The "if" part is obvious. Assume therefore that $\sigma(v, v) \neq 0$ for all non-zero vectors v.

If there exist vectors a and b such that $\sigma(a, a) > 0$ and $\sigma(b, b) < 0$, then $\sigma(a, b)^2 - \sigma(a, a)\sigma(b, b) > 0$ and therefore the equation

$$\sigma(Xa - b, Xa - b) = 0,$$

i.e.,

$$\sigma(a, a)X^2 - 2\sigma(a, b)X + \sigma(b, b) = 0,$$

has real roots. If x is one such root, then we must have $b = xa$, whence $\sigma(b, b) = x^2\sigma(a, a) \geq 0$, which is a contradiction. Hence σ is either positive definite or negative definite.

It follows from Proposition 2 above and Lemma 4 of Chapter V (p. 97) that a definite polarity $\perp$ determines a unique similarity class of distances D. Hence our *similarity euclidean geometry* (A, D) *is determined in* $\mathscr{P}(V)$ *by H and the definite polarity $\perp$ of* $\mathscr{P}(H)$. We also use the notation (A, $\perp$) for (A, D). Moreover, every similarity euclidean geometry is isomorphic to such a geometry (A, $\perp$).

We can define in (A, D) ratios of distances by means of the class D. We may also define angles in exactly the same way as for euclidean geometry since the expression for $\cos \theta$ is unaltered when d is multiplied by an arbitrary positive scalar. Both these notions may be interpreted in $\mathscr{P}(V)$ by using the so-called cross-ratios (cf. exercises 5, 6 below).

Suppose that L, L' are lines in (A, $\perp$) and that the angle between them is $\pi/2$. This case may be interpreted without cross-ratios. By the definition given on page 128, this means that the points at infinity on the two lines are orthogonal with respect to $\perp$, i.e., $(L \cap H)^{\perp} \supset L' \cap H$. More generally, we call the elements M, M' of $(\mathsf{A}, \perp)$ *perpendicular* if one of $M \cap H$, $(M' \cap H)^{\perp}$ contains the other.

EXERCISES

1. Let A_0, A_1, A_2, A_3 be distinct collinear points in a projective geometry $\mathscr{P}(V)$ over an arbitrary field F. Choose homogeneous vectors such that $[a_0] = A_0$, $[a_1] = A_1$, $[a_0 + a_1] = A_2$ and $[xa_0 + a_1] = A_3$. Show that x is uniquely determined by the points A_0, A_1, A_2, A_3 (in that order). The scalar x is called the *cross-ratio* of the quadruple and will be written $\mathrm{cr}(A_0, A_1; A_2, A_3)$.

   If π is a collineation of $\mathscr{P}(V)$ prove that

   $$\mathrm{cr}(A_0\pi, A_1\pi; A_2\pi, A_3\pi) = \mathrm{cr}(A_0, A_1; A_2, A_3)$$

   and deduce that if $B_i = [y_i a_0 + z_i a_1]$, $i = 0, 1, 2, 3$, are distinct points, then

   $$\mathrm{cr}(B_0, B_1; B_2, B_3) = \frac{(y_0 z_2 - y_2 z_0)(y_1 z_3 - y_3 z_1)}{(y_0 z_3 - y_3 z_0)(y_1 z_2 - y_2 z_1)}.$$

   When F is not of characteristic 2 show that

   $$\mathrm{cr}(A_0, A_1; A_2, A_3) = \mathrm{cr}(A_0, A_1; A_3, A_2)$$

   if, and only if,

   $$\mathrm{cr}(A_0, A_1; A_2, A_3) = -1$$

   and that this is precisely the condition for $(A_0, A_1; A_2, A_3)$ to be a harmonic range (see p. 40).

2. Let $\mathscr{P}(V)$ be a real projective geometry and (A, D) the similarity euclidean geometry in $\mathscr{P}(V)$ determined by H and the distance class D on $\mathscr{A}(H)$. If A_1, A_2, A_3 are distinct collinear points in (A, D) and A_0 is the point at infinity on the line $A_1 A_2$, prove that

   $$|\mathrm{cr}(A_0, A_1; A_2, A_3)| = d(A_1, A_3)/d(A_1, A_2)$$

   for every d in D.

3. Let r be a given positive real number and C a point in a euclidean geometry $\mathscr{A}(S, d)$. We define the *sphere* with *center* C of *radius* r to be the set of all points P satisfying $d(P, C) = r$. Find the equation of this sphere with respect to a euclidean coordinate system and deduce that it is an affine quadric. Show also that the quadric is non-degenerate and that its center is C.

   If d' is a distance similar to d, then the quadric is still a sphere with center C in $\mathscr{A}(S, d')$ but possibly of different radius. We may therefore define a set of points in a similarity euclidean geometry $\mathscr{A}(S, D)$ to be a sphere if the set is a sphere in $\mathscr{A}(S, d)$ for any d in D.

4. Let $(\mathsf{A}, \perp)$ be the similarity euclidean geometry determined in $\mathscr{P}(V)$ by H and the definite polarity $\perp = \perp(\sigma)$. If $Q(\tau)$ is a projective quadric prove that $Q(\tau) \cap \mathsf{A}$ is a sphere if, and only if, $Q(\tau)$ contains more than one point and $\perp(\tau_H) = \perp(\sigma)$.

5. Let $(A, \perp)$ be a similarity euclidean geometry as in exercise 4. Given points A_1, A_2, B_1, B_2, where $A_1 \neq A_2$ and $B_1 \neq B_2$, describe a purely projective geometrical method of calculating the ratio $d(A_1, A_2)/d(B_1, B_2)$. (Use exercises 2 and 4.)

6. Let H be a line in a real projective plane $\mathscr{P}(V)$ and $\perp(\sigma)$ a definite polarity in $\mathscr{P}(H)$. Complexify V and extend σ to σ' as in § 5.6 (p. 113). Then the quadric $Q(\sigma')$ consists of two points I and J. (These are called the "circular points at infinity" for the similarity euclidean plane $(A, \perp)$.)

   Given a point A in $\mathscr{P}(V)$ but not in H, show that a coordinate system of $\mathscr{P}(V)$ may be chosen so that A has coordinates $(1, 0, 0)$ and in the corresponding coordinate system of $\mathscr{P}(V_{(\mathbb{C})})$, I and J have coordinates $(0, 1, i)$, $(0, 1, -i)$, respectively.

   Using this coordinate system, prove the following result: If B and C are points of $\mathscr{P}(V)$ distinct from each other and from A, let $B' = AB \cap H$ and $C' = AC \cap H$. Then the real part of $\mathrm{cr}(I, J; B', C')$ is $\cos 2\theta$, where θ is the angle $\widehat{BAC}$ in the similarity euclidean plane $(A, \perp)$.

7. Let $(A, \perp)$ be the 3-dimensional similarity euclidean geometry determined by H and $\perp$ in $\mathscr{P}(V)$. Given a plane M and a point P not in M, prove that there is a unique line through P perpendicular to M. This line is called the *normal* to M through P. Show that all the normals to M are parallel.

   If M' is a second plane, then M' is perpendicular to M if, and only if, every normal to M is perpendicular to every normal to M'.

6.3   Euclidean Quadrics

We call an affine quadric in a euclidean geometry a *euclidean quadric*. In § 5.7 (Proposition 12) we saw how to choose an affine coordinate system for which the equation of a given affine quadric reduces to one of two simple types. In this section we shall show that the same reduction can be achieved by a suitable choice of *euclidean* coordinate system.

We begin by considering two symmetric bilinear forms σ, τ on the same real vector space V and suppose that σ is positive definite. (In the discussion up to Lemma 2 the ground field could be quite arbitrary and we need only assume that σ is non-degenerate.) Our first aim is to find a basis of V whose vectors are mutually orthogonal with respect to σ and τ simultaneously.

We may define unambiguously a linear mapping f of V into V by means of the equation

$$\sigma(af, b) = \tau(a, b) \tag{1}$$

for all a, b in V. This may be seen quite simply in terms of matrices: for if S, T are the matrices of σ, τ, respectively, with respect to an ordered basis $(v_1, \ldots, v_n)$ of V, then the matrix A of f must satisfy

$$A S = T \quad \text{or} \quad A = T S^{-1}$$

because S is invertible. Alternatively we may note that, as σ is non-degenerate, the mapping g defined on p. 91 is an isomorphism and $f = \tau g^{-1}$. From the symmetry of σ and τ we deduce

$$\sigma(af, b) = \sigma(a, bf) \tag{2}$$

for all a, b in V.

Conversely, if a symmetric bilinear form σ and a linear mapping f satisfying equation (2) are given, then we may define a symmetric bilinear form τ on V by means of equation (1).

DEFINITION. A linear mapping f satisfying equation (2) is called σ-symmetric.

Since (V, σ) is a euclidean space we observe that a linear mapping f is σ-symmetric if, and only if, it has a symmetric matrix with respect to any cartesian basis of (V, σ). (See also exercise 1 below.)

It is helpful to interpret the mapping f in terms of the projective geometry $\mathscr{P}(V)$. When σ and τ are both non-degenerate they determine polarities whose product $\perp(\tau)\perp(\sigma)$ is a collineation which is precisely $\mathscr{P}(f)$. Our problem is to find a simplex which is self-polar with respect to $\perp(\sigma)$ and $\perp(\tau)$ simultaneously. (Cf. exercise 15 of § 5.6.) The vertices of such a simplex must be fixed points of $\mathscr{P}(f)$. This leads to the following concept (cf. exercise 4 of § 4.2).

DEFINITION. Let V be a vector space over an arbitrary field F and f a linear mapping of V into V. A non-zero vector a of V is an *eigenvector* of f if $[a]f \subset [a]$. In other words, a is an eigenvector of f if $a \neq 0$ and there exists a scalar x in F such that $af = xa$. The scalar x corresponding to such a vector a is called an *eigenvalue* of f. We call $\mathrm{Ker}(f - x1)$ the *eigenspace* of f corresponding to x and shall write this as $E_x(f)$ or as E_x when f is understood.

LEMMA 2. *Let (V, σ) be a (non-zero) euclidean space. If f is a σ-symmetric mapping of V into V then f has eigenvectors in V.*

Our proof depends on the following important general fact.

PROPOSITION 3. *If f is a linear mapping of the (non-zero) complex vector space W into itself, then f has eigenvectors in W.*

PROOF. Since W is finite dimensional there is a least positive integer r such that f^r is linearly dependent on $f^{r-1}, \ldots, f, 1_W$, say

$$f^r = c_1 f^{r-1} + \cdots + c_r 1_W.$$

If we put

$$m(X) = X^r - c_1 X^{r-1} - \cdots - c_r$$

then we may express this dependence by writing

$$m(f) = 0.$$

The equation $m(X) = 0$ always has a complex root x and we have the corresponding factorization

$$m(X) = (X - x)\, n(X) \tag{3}$$

in $\mathbf{C}[X]$. Substituting f in equation (3) and using rules A.1 to A.4 in the linear algebra $\mathscr{L}(W, W)$ (cf. p. 68), we obtain

$$(f - x1)n(f) = 0.$$

Now $f - x1$ cannot be an isomorphism as otherwise $n(f) = 0$. This would contradict the minimal character of the degree of $m(X)$. Hence there exists a non-zero vector c in W such that $cf = xc$.

PROOF OF LEMMA 2. Extend f to the complexification $V_{(C)}$ of V by the rule $(u + iv)f = uf + i(vf)$. Clearly the extended f is a linear mapping of $V_{(C)}$ into $V_{(C)}$. By Proposition 3 there exists an eigenvector $c = a + ib$ with eigenvalue x. We have to show that x is real.

Consider the extended bilinear form σ' on $V_{(C)}$ as defined in § 5.6 (p. 113). The equation

$$\sigma'(wf, w') = \sigma'(w, w'f)$$

is now true for all w, w' in $V_{(C)}$. Hence, in particular,

$$\sigma'(\bar{c}f, c) = \sigma'(\bar{c}, cf)$$

where $\bar{c} = a - ib$. But $cf = xc$ and so $\bar{c}f = \bar{x}\bar{c}$ (where $\bar{x}$ denotes the complex conjugate of x). Therefore

$$\bar{x}\sigma'(\bar{c}, c) = x\sigma'(\bar{c}, c),$$

where $\sigma'(\bar{c}, c) = \sigma(a, a) + \sigma(b, b) > 0$. This implies that x is real.

The equation $cf = xc$ now gives $af = xa$ and $bf = xb$. Since c is non-zero, either a or b is an eigenvector of f in V.

LEMMA 3. *If f is a σ-symmetric mapping on V, then*

$$[a]f \subset [a] \quad \text{implies} \quad [a]^{\perp}f \subset [a]^{\perp}.$$

PROOF. If $b \in [a]^{\perp}$, then

$$\sigma(a, bf) = \sigma(af, b) = x\sigma(a, b) = 0$$

and so $bf \in [a]^{\perp}$.

PROPOSITION 4. *If (V, σ) is a euclidean space and f is a σ-symmetric linear mapping of V into V then there is a cartesian basis of (V, σ) consisting entirely of eigenvectors of f.*

PROOF. By Lemma 2 we can find an eigenvector a_1 of f. Since σ is positive definite we may choose a_1 so that $\sigma(a_1, a_1) = 1$. The subspace $[a_1]$ is non-degenerate with respect to σ and therefore

$$V = [a_1] \oplus [a_1]^{\perp(\sigma)}$$

(by Proposition 5, Chapter V, p. 96). Our proposition now follows by Lemma 3 and induction on the dimension of V.

THEOREM 1. *If σ and τ are symmetric bilinear forms on the real vector space V and σ is positive definite, then there exists a cartesian basis of (V, σ) whose vectors are mutually orthogonal with respect to τ.*

PROOF. Let f be the σ-symmetric linear mapping determined by σ, τ according to equation (1) above. Then by Proposition 4, (V, σ) has a cartesian basis $(a_1, \ldots, a_n)$ such that

$$a_i f = x_i a_i \quad \text{for} \quad i = 1, \ldots, n.$$

Thus

$$\tau(a_i, a_j) = \sigma(a_i f, a_j) = x_i \sigma(a_i, a_j) = x_i \delta_{ij},$$

and so $a_1, \ldots, a_n$ are mutually orthogonal with respect to τ.

As we might expect, Theorem 1 may be expressed in terms of matrices. Recall from p. 134 that if σ, τ have matrices S, T, respectively, then f has matrix $A = T S^{-1}$. The cartesian basis vectors of Theorem 1 may be obtained by finding their coordinate rows c_i which satisfy

$$c_i A = x_i c_i$$

or equivalently

$$c_i T = x_i c_i S.$$

The scalars x_i can be found by using the routine method of solution given for linear equations in § 3.3. Those who are familiar with determinants will recognize that the scalars x_i are also given by the condition

$$\det (T - x_i S) = 0$$

or

$$\det (A - x_i I) = 0.$$

They are naturally called the *eigenvalues* of the matrix A.

As in the proof of Theorem 1, the rows c_i satisfy

$$c_i S c_j{}^t = \delta_{ij} \quad \text{and} \quad c_i T c_j{}^t = x_i \delta_{ij}.$$

In other words, if the matrix P has rows $c_1, \ldots, c_n$, then

$$P S P^t = I \quad \text{and} \quad P T P^t = \text{diag}(x_1, \ldots, x_n)$$

(see p. 104). Thus *Theorem 1 may be regarded as a result about the simultaneous reduction of two matrices to diagonal form.*

As an illustration of these comments consider the symmetric matrices

$$S = \begin{pmatrix} 2 & 2 & 1 \\ 2 & 5 & 2 \\ 1 & 2 & 1 \end{pmatrix}, \qquad T = \begin{pmatrix} 3 & 4 & 2 \\ 4 & 9 & 4 \\ 2 & 4 & 2 \end{pmatrix}.$$

The required eigenvalues are the roots of the equation

$$\begin{vmatrix} 3-2X & 4-2X & 2-X \\ 4-2X & 9-5X & 4-2X \\ 2-X & 4-2X & 2-X \end{vmatrix} = 0,$$

which, by simple row operations, gives

$$\begin{vmatrix} 1-X & 0 & 0 \\ 0 & 1-X & 0 \\ 2-X & 4-2X & 2-X \end{vmatrix} = 0,$$

i.e., $(1-X)^2(2-X)=0$.

Corresponding to the root 1 the matrix equation $(X_1, X_2, X_3)(T-S)$ $=0$ yields $X_1+2X_2+X_3=0$. One solution is $d_1=(1, -1, 1)$. We check that $d_1 S\, d_1{}^t = 2$ and so we "normalize" d_1 by dividing by $\sqrt{2}$ and setting $c_1 = \dfrac{1}{\sqrt{2}}(1, -1, 1)$.

We require another solution d_2 which is also S-orthogonal to d_1, i.e., $d_2 S\, d_1{}^t = 0$. Thus d_2 satisfies $X_1+2X_2+X_3=0$ and $X_1-X_2=0$. We may take $d_2 = (1, 1, -3)$. As $d_2 S\, d_2{}^t = 2$ we set $c_2 = \dfrac{1}{\sqrt{2}}(1, 1, -3)$.

Corresponding to the root 2 we solve $(X_1, X_2, X_3)(T-2S)=0$, i.e., $-X_1=0$ and $-X_2=0$. So we may take $c_3 = (0, 0, 1)$.

It follows as above that if

$$P = \begin{pmatrix} 1/\sqrt{2} & -1/\sqrt{2} & 1/\sqrt{2} \\ 1/\sqrt{2} & 1/\sqrt{2} & -3/\sqrt{2} \\ 0 & 0 & 1 \end{pmatrix},$$

then $P S\, P^t = I$ and $P T\, P^t = \mathrm{diag}\,(1, 1, 2)$.

THEOREM 2. *Given a euclidean quadric in an n-dimensional euclidean geometry, then a euclidean coordinate system can be chosen with respect to which the quadric has an equation of one of the two types:*

$$d_1 X_1{}^2 + \cdots + d_n X_n{}^2 + d_0 = 0 \tag{4}$$
$$X_1 + d_2 X_2{}^2 + \cdots + d_n X_n{}^2 = 0 \tag{5}$$

(where some, or all, of the d_i's may be zero).

PROOF. Our theorem differs from Proposition 12 of Chapter V only in the fact that the required coordinate system must be euclidean. Our proof is a refinement of the proof of the earlier proposition.

Let $\mathscr{P}(V)$ be an n-dimensional real projective geometry, H a hyperplane in $\mathscr{P}(V)$ and σ a positive definite bilinear form on H. We may suppose our euclidean geometry to be $\mathscr{A}(c+H, d)$ for some c not in H and d the distance determined by σ. The given quadric, say R, is the cross-section by $\mathscr{A}(c+H)$ of a suitable projective quadric $Q(\tau)$ in $\mathscr{P}(V)$. (Cf. p. 116.)

Our aim is to find an ordered basis $(a_0, \ldots, a_n)$ of V adapted to τ and such that $a_0 \in c+H$ and $(a_1, \ldots, a_n)$ is a cartesian basis of (H, σ). For then $a_0+a_1, \ldots, a_0+a_n, a_0$ will be vectors for a euclidean frame of reference of $\mathscr{A}(c+H, d)$. If $f(X_0, \ldots, X_n) = 0$ is the equation of $Q(\tau)$ then $f(1, X_1, \ldots, X_n) = 0$ is the equation of R with respect to this euclidean frame of reference.

In the following argument we shall write $\perp$ for $\perp(\tau)$.

(i) If $H^\perp \not\subset H$, we may find a vector a_0 in $H^\perp \cap c+H$. Choose a cartesian basis $(a_1, \ldots, a_n)$ of (H, σ) with the property that $\tau(a_i, a_j) = 0$ whenever $i \neq j$ $(i, j = 1, \ldots, n)$. (Such a basis exists by Theorem 1.) The equation of Q is then of the form

$$d_0 X_0{}^2 + \cdots + d_n X_n{}^2 = 0.$$

Substituting $X_0 = 1$ yields the equation of R in the form (4).

(ii) If $H^\perp \subset H$, then $V^\perp \subset H$ and $\dim V^\perp = \dim H^\perp - 1$ (cf. the proof of Proposition 12, p. 118). Let $(a_{r+1}, \ldots, a_n)$ be a cartesian basis of $V^\perp$. By Proposition 5 of Chapter V (p. 96) we can, first of all, extend this to a cartesian basis $(a_1, a_{r+1}, \ldots, a_n)$ of $H^\perp$ and then write $H = H^\perp \oplus M$, where $M = (H^\perp)^{\perp(\sigma)}$. By Theorem 1 above, M has a cartesian basis $(a_2, \ldots, a_r)$ whose vectors are mutually orthogonal with respect to τ. Therefore $(a_1, \ldots, a_n)$ is a cartesian basis of H such that $\tau(a_i, a_j) = 0$ whenever $i \neq j$ $(i, j = 1, \ldots, n)$.

Let $K = [a_2, \ldots, a_n]$. We assert that $K^\perp$ is not contained in H. For $K = V^\perp \oplus M$, so that $K^\perp = M^\perp$. On the other hand, $H = H^\perp \oplus M$ implies that τ_M is non-degenerate (Lemma 3, Corollary, p. 95) and so $V = M \oplus M^\perp$ (Proposition 5, p. 96).

Choose a line L in $K^\perp$ containing $[a_1]$ and not contained in H. Then L meets Q in one further point $[a_0]$, where $a_0 \in c+H$. The equation of Q in the coordinate system determined by $(a_0, \ldots, a_n)$ is

$$X_0 X_1 + d_2 X_2{}^2 + \cdots + d_n X_n{}^2 = 0,$$

for suitable scalars $d_2, \ldots, d_n$. Putting $X_0 = 1$ yields the equation of R in the form (5).

(To prove (ii) the reader may find it more instructive to consider first the case where τ is non-degenerate. For then $[a_1] = H^{\perp(\tau)}$ and $K = [a_1]^{\perp(\sigma)}$ are uniquely determined once τ is given. In the general case one can find a subspace N which has $(V^{\perp(\tau)})^{\perp(\sigma)}$ as hyperplane at infinity; then $V = V^{\perp(\tau)} \oplus N$ and one can treat the restriction of τ to N as above.)

EXERCISES

1. Suppose σ is a symmetric bilinear form on a vector space V and f a linear mapping of V into V. If $(a_1, \ldots, a_n)$ is an ordered basis of V and S, A are the matrices of σ, f with respect to this basis, prove that f is σ-symmetric if, and only if, $AS = SA^t$.

2. If a is a given non-zero vector in a complex vector space V and f is a linear mapping of V into V, prove that there exists a polynomial $p(X)$ such that $ap(f)$ is an eigenvector of f.
   (Consider the vectors af^i, $i = 0, 1, 2, \ldots$ and proceed as in the proof of Proposition 3.)

3. If V is a complex vector space and f, g are commuting linear mappings of V into V (i.e., $fg = gf$), use exercise 2 to show that there exists a common eigenvector for f and g. Show that there exists a basis with respect to which the matrices of f and g are both triangular.

4. Let σ, τ be symmetric bilinear forms on a real vector space V and let σ be positive definite. Assume the theorem of analysis that the real valued (continuous) function $a \rightarrow \tau(a, a)$ $(a \in V)$ attains a maximum x_1 at some point a_1 on the (compact) sphere $\sigma(a, a) = 1$. Show that $x_1\sigma - \tau$ is positive semi-definite and use exercise 3 of §5.4 to prove that $[a_1]^{\perp(\sigma)} \subset [a_1]^{\perp(\tau)}$. Hence prove Theorem 1.

6.4 Euclidean Automorphisms

It follows from Proposition 1 that an automorphism of a euclidean geometry $\mathscr{A}(a + M, d)$ (i.e., an isomorphism of the geometry onto itself) is an affinity whose unique underlying linear automorphism is an automorphism of the euclidean space (M, d). Our aim is now to determine the structure of such an automorphism of (M, d).

We prefer to change the notation and work with a euclidean space (V, σ). As usual, $V_{(\mathbb{C})}$ denotes the complexification of V and as in §5.6 (p. 113) we extend σ to a bilinear form σ' on $V_{(\mathbb{C})}$. If $c = a + ib$, we shall write $\bar{c} = a - ib$ and of course use the bar over scalars to denote complex conjugation. We observe that $\sigma'(\bar{c}, c) = \sigma(a, a) + \sigma(b, b)$ and thus $\sigma'(\bar{c}, c) > 0$ for all $c \neq 0$. (Cf. §6.5, p. 146.)

Let f be an automorphism of (V, σ): so f is a one–one linear mapping of V onto V such that

$$\sigma(af, bf) = \sigma(a, b)$$

for all a, b in V. We extend f to $V_{(C)}$ by the rule

$$(a+ib)f = af+i(bf)$$

and leave the reader to check that the extended f is an automorphism of the bilinear space $(V_{(C)}, \sigma')$.

LEMMA 4. *Every eigenvalue of f is of the form $e^{i\theta}$ for some real θ satisfying $-\pi < \theta \leq \pi$.*

PROOF. Let $c = a+ib$ be an eigenvector of f with given eigenvalue x: $cf = xc$. Then $\bar{c}f = \bar{x}\bar{c}$ and hence

$$\sigma'(\bar{c}f, cf) = \sigma'(\bar{x}\bar{c}, xc) = \bar{x}x\sigma'(\bar{c}, c);$$

while

$$\sigma'(\bar{c}f, cf) = \sigma'(\bar{c}, c)$$

because f preserves σ'. Since $c \neq 0$, $\sigma'(\bar{c}, c) > 0$ and hence $\bar{x}x = 1$. Thus $x = e^{i\theta}$ for some real number θ which may be chosen to satisfy $-\pi < \theta \leq \pi$.

LEMMA 5. *To each eigenvector c of f with non-real eigenvalue $e^{i\theta}$ there corresponds a two-dimensional subspace T of V such that $Tf = T$ and a cartesian basis (u, v) of (T, σ_T) such that*

$$(f; (u, v)) = \begin{pmatrix} \cos\theta & -\sin\theta \\ \sin\theta & \cos\theta \end{pmatrix}.$$

PROOF. Let $c = a+ib$. Then

$$\sigma'(c, c) = \sigma'(cf, cf) = e^{2i\theta}\sigma'(c, c)$$

and therefore $\sigma'(c, c) = 0$ because $e^{2i\theta} \neq 1$ (as $e^{i\theta}$ is not real). But

$$\sigma'(c, c) = \sigma(a, a) - \sigma(b, b) + 2i\sigma(a, b)$$

whence

$$\sigma(a, a) = \sigma(b, b)$$

and

$$\sigma(a, b) = 0.$$

From this last equation it follows that a and b are linearly independent. Now if $u = ra$, $v = rb$ where $r^2\sigma(a, a) = 1$, and T is the subspace spanned by u, v, then (u, v) is a cartesian basis of (T, σ_T). Finally,

$$(rc)f = e^{i\theta}(rc)$$

gives

$$uf + i(vf) = (\cos\theta + i\sin\theta)(u+iv)$$

which shows that $Tf = T$ and yields the required matrix of f with respect to (u, v).

LEMMA 6. *If $Mf = M$, then $(M^{\perp})f = M^{\perp}$.*

PROOF. For any a in M and b in $M^{\perp}$ we have $af^{-1} \in M$ and so

$$\sigma(a, bf) = \sigma(af^{-1}, b) = 0.$$

Thus $(M^{\perp})f \subset M^{\perp}$ and equality follows because f is one–one.

THEOREM 3. *Let f be an automorphism of the euclidean space (V, σ). Then there exists a cartesian basis of (V, σ) with respect to which f has matrix*

$$\begin{pmatrix} A_1 & & & & & \\ & A_2 & & & & \\ & & \ddots & & & \\ & & & A_r & & \\ & & & & I_s & \\ & & & & & -I_t \end{pmatrix}$$

where

$$A_k = \begin{pmatrix} \cos \theta_k & -\sin \theta_k \\ \sin \theta_k & \cos \theta_k \end{pmatrix}$$

$(0 < |\theta_k| < \pi)$ *for $k = 1, \ldots, r$. The integers r, s, t and also the real numbers $|\theta_1|, \ldots, |\theta_r|$ (not necessarily distinct) are invariants of f.*

PROOF. Let c be an eigenvector of f with eigenvalue $e^{i\theta}$. If $e^{i\theta}$ is real then we set $M = [c]$. If $e^{i\theta}$ is not real then we set $M = T$ as in Lemma 5. In either case $V = M \oplus M^{\perp}$ (by Proposition 5 of Chapter V, p. 96) and $Mf = M$, $M^{\perp}f = M^{\perp}$ (by Lemma 6). The existence of a cartesian basis as required for our theorem now follows by induction on dim V.

We recall from p. 135 the definition of the eigenspace E_x as $\mathrm{Ker}\,(f - x1)$. It is clear that $s = \dim E_1$ and $t = \dim E_{-1}$ so that s and t are invariants of f. If f now denotes the extended automorphism of $V_{(\mathbf{C})}$ and $e^{i\theta}$ is a non-real eigenvalue of f then $|\theta|$ occurs among $|\theta_1|, \ldots, |\theta_r|$ and is repeated exactly m times where m is the dimension over $\mathbf{C}$ of the eigenspace $E_{e^{i\theta}}$ in $V_{(\mathbf{C})}$.

Let f be an arbitrary linear mapping of V into V and A the matrix of f with respect to a cartesian basis $(a_1, \ldots, a_n)$ of (V, σ). Then f satisfies the condition

$$\sigma(af, bf) = \sigma(a, b)$$

for all a, b in V (and hence is one–one) if, and only if,

$$A A^t = I_n$$

i.e., if, and only if, A is *orthogonal*. (Cf. exercise 2 of § 4.4, p. 79.)

Theorem 3 therefore yields the following

COROLLARY. *If A is an orthogonal matrix, then there exists another orthogonal matrix P such that $P^{-1}A\,P$ is of the form displayed in Theorem 3.*

PROOF. Let f be the automorphism of (V, σ) whose matrix with respect to the cartesian basis $(a_1, \ldots, a_n)$ is A. By Theorem 3 there exists a cartesian basis $(b_1, \ldots, b_n)$ such that $B = (f; (b_i))$ has the required form. Then $B = P^{-1}A\,P$ where $P = (1_V; (a_i), (b_j))$.

Let us now consider the special case of Theorem 3 when dim $V = 2$. Here we have two possibilities:

(i) the matrix of f is

$$\begin{pmatrix} \cos\theta & -\sin\theta \\ \sin\theta & \cos\theta \end{pmatrix}$$

where $-\pi < \theta \leq \pi$; or

(ii) the matrix of f is

$$\begin{pmatrix} x & 0 \\ 0 & y \end{pmatrix}$$

where $x = +1$, $y = -1$ or $x = -1$, $y = +1$.

In case (i), if we take any non-zero vector a and calculate the angle φ between a and af, we find

$$\cos\varphi = \frac{\sigma(a, af)}{\sigma(a, a)} = \cos\theta.$$

Since $\varphi \in [0, \pi]$ by definition,

$$\varphi = \theta \quad \text{if} \quad \theta \geq 0$$
$$\varphi = -\theta \quad \text{if} \quad \theta < 0.$$

In any event, the angle between any non-zero vector and its image is constant. We call f a *rotation through the angle* θ. Note that the automorphisms 1_V and -1_V are included as rotations (when θ equals 0 and π, respectively). In case (ii) f is said to be a *reflexion* ("in the X-axis" if $y = -1$, "in the Y-axis" if $x = -1$).

These considerations suggest the following

DEFINITION. Let f be an automorphism of the euclidean space (V, σ). If E_{-1}, the eigenspace corresponding to -1, has even dimension we call f a *rotation*; while if E_{-1} has odd dimension we say f is a *reflexion*.

Let us assume for the moment an elementary knowledge of determinants. Theorem 3 shows that $\det f$, the determinant of f, is $+1$ or -1 according as $\dim E_{-1}$ is even or odd. Thus f is a rotation if, and only if, $\det f = +1$.

Consider now the set of all automorphisms of (V, σ). This is obviously a subgroup (with respect to multiplication) of Aut V. It will be denoted by Aut(V, σ) and is called the *orthogonal group* on (V, σ).

The mapping $f \to \det f$ of Aut(V, σ) into $\mathbf{R}^*$ (actually onto $\{+1, -1\}$) is a homomorphism of groups. The kernel is precisely the set of all rotations and this is therefore a subgroup, called the *rotation group* on (V, σ). We say that two cartesian bases (a_i), (b_i) of (V, σ) have the same *orientation* if the euclidean automorphism $a_i \to b_i$, for all i, is a rotation. In view of the fact that the rotations form a group, this defines an equivalence relation on the set of all cartesian bases of (V, σ). There are, of course, just two equivalence classes. We *orient* (V, σ) by assigning to the space one of these two classes.

Suppose we now form the similarity class of distances $D = (d)$ and look at the set of all f in Aut V that preserve D (in the natural sense that to every d in D there corresponds d' in D such that $d(af, bf) = d'(a, b)$ for all a, b in V). This set is easily seen to be a subgroup of Aut V and will be written Aut(V, D).

Our definition of similarity euclidean isomorphism and Proposition 6 of Chapter III, p. 61, immediately show that Aut $\mathscr{A}(V, D)$, the group of all similarity euclidean automorphisms of $\mathscr{A}(V, D)$, is the split product of (G, T), where T is the group of all translations and G is isomorphic to Aut(V, D).

If $f \in$ Aut (V, D) and $d \in D$, then

$$d(af, bf) = r\, d(a, b)$$

for some uniquely determined positive real number r. This shows that $(1/r)f \in$ Aut(V, d) and thus Aut(V, D) is the split product of $(\text{Aut}(V, d), P)$, where P is the group of all linear automorphisms $r1$ with r a positive real number.

Finally we propose to describe the automorphism group of a similarity euclidean geometry in projective geometrical terms. We shall use the following

DEFINITION. If $\mathscr{P}(V)$ is a projective geometry over an arbitrary field, $\perp(\sigma)$ is a polarity of $\mathscr{P}(V)$ and $\mathscr{P}(f)$ is a collineation such that

$$\perp(\sigma)\mathscr{P}(f) = \mathscr{P}(f)\perp(\sigma)$$

then $\mathscr{P}(f)$ is said to *preserve* $\perp(\sigma)$.

LEMMA 7. *Let $\perp$ be a definite polarity of the real projective geometry $\mathscr{P}(H)$ and let D be the corresponding distance class on H. Then a collineation $\mathscr{P}(g)$ of $\mathscr{P}(H)$ preserves $\perp$ if, and only if, g preserves D.*

PROOF. We recall the notation σ^g introduced on p. 89. We also use the notation d^g in the same way and denote by D^g the set of all d^g for d in D.

If M is any subspace of H, then $M^{\perp(\sigma^g)}$ is the set of all vectors b such that $\sigma^g(M, b) = 0$. Hence $M^{\perp(\sigma^g)}g$ is the set of all vectors bg such that $\sigma(Mg, bg) = 0$. This means that $M^{\perp(\sigma^g)}g = (Mg)^{\perp(\sigma)}$. We thus have the general formula

$$\perp(\sigma^g)\mathscr{P}(g) = \mathscr{P}(g)\perp(\sigma).$$

We conclude that $\mathscr{P}(g)$ preserves $\perp(\sigma)$ if, and only if, $\perp(\sigma^g) = \perp(\sigma)$. This in turn is equivalent to $D^g = D$, as required.

PROPOSITION 5. *Let $(A, \perp)$ be the similarity euclidean geometry in $\mathscr{P}(V)$ determined by H and the definite polarity $\perp$ of $\mathscr{P}(H)$. Denote by $(\mathrm{Pr}\,\mathscr{P}(V))_{H,\perp}$ the subgroup of the projective group $\mathrm{Pr}\,\mathscr{P}(V)$ whose elements restrict to collineations of $\mathscr{P}(H)$ preserving $\perp$. Then the isomorphism of $(\mathrm{Pr}\,\mathscr{P}(V))_H$ onto $\mathrm{Af}\ A$ (defined in Theorem 7 of Chapter III, p. 61) restricts to an isomorphism of $(\mathrm{Pr}\,\mathscr{P}(V))_{H,\perp}$ onto $\mathrm{Aut}(A, \perp)$.*

PROOF. Let $\mathscr{P}(f)$ be an element of $(\mathrm{Pr}\,\mathscr{P}(V))_H$ and let g be the restriction of f to H. By Lemma 7, $\mathscr{P}(g)$ preserves $\perp$ if, and only if, g preserves D. By our definition of the distance class D on A this means precisely that the restriction of $\mathscr{P}(f)$ to A is an automorphism of (A, D).

EXERCISES

1. Show that $+1$ is always an eigenvalue of a rotation in an odd dimensional euclidean space.

2. If L and L' are two non-parallel lines in a euclidean plane $\mathscr{A}(V, d)$ show that there are exactly four automorphisms which leave the point $L \cap L'$ fixed and take L onto L'.

   Show that two of the underlying automorphisms of the euclidean space (V, d) are rotations and compare their angles.

3. Exhibit a homomorphism of the additive group **R** onto the group of rotations of a 2-dimensional euclidean space and identify the kernel of your homomorphism.

4. Show that a reflexion of a 2-dimensional euclidean space has a line of fixed points. Show that a reflexion of a 3-dimensional euclidean space which has a fixed point other than the origin, must have a whole plane of fixed points.

6.5   Hilbert Spaces

DEFINITION.   If W is a (finite dimensional) complex vector space and σ a positive definite (hermitian) form on W, then (W, σ) is called a *Hilbert space*.

Hilbert spaces arise naturally from euclidean spaces: for whenever (V, σ) is a euclidean space and we construct the complexification $\sigma_{(C)}$ of σ (cf. p. 124) then $(V_{(C)}, \sigma_{(C)})$ is a Hilbert space.

We have already used Hilbert spaces in the last two sections.   For whenever we used the bilinear extension σ' of σ to $V_{(C)}$ we could have employed $\sigma_{(C)}$ in a suitably modified, but essentially similar, argument. There is a precise connexion between σ' and $\sigma_{(C)}$.   If $v = a + ib$, let us denote by $\bar{v}$ the vector $a - ib$.   Then $v \to \bar{v}$ is a semi-linear automorphism of $V_{(C)}$ and $\sigma_{(C)}(u, v) = \sigma'(\bar{u}, v)$ for all u, v in $V_{(C)}$.   If f is an automorphism of (V, σ), then the natural extension of f:

$$(a + ib)f = af + ibf,$$

is an automorphism of $(V_{(C)}, \sigma_{(C)})$ (cf. exercise 4 of § 5.8, p. 124).

Let (W, σ) be a Hilbert space.   For a in W we may define $\|a\|$ (the length or norm of a) and $d(a, b)$ (the distance from a to b) exactly as in a euclidean space.   The rules D.1–D.6 all hold but, of course, in D.5 we take x in C and interpret $|x|$ as $+\sqrt{x\bar{x}}$.   Every mapping d satisfying D.1–D.6 arises in this way (cf. exercise 2 below).

We have already remarked in § 5.8 (p. 123) that W possesses an ordered basis $(a_1, \ldots, a_n)$ with respect to which the matrix of σ is I_n. As in the euclidean case, we call $(a_1, \ldots, a_n)$ a *cartesian basis* of (W, σ). A linear automorphism f of W is an automorphism of (W, σ) if, and only if, the matrix P of f with respect to a cartesian basis satisfies $P \bar{P}^t = I_n$.   (Such a matrix is called *unitary*.)   The set of all automorphisms of (W, σ) forms the *unitary group* $\text{Aut}(W, \sigma)$.

Both Theorems 1 and 3 have natural complex analogues.   We state these explicitly.

THEOREM 1H.   *If σ, τ are hermitian forms on the complex vector space W and σ is positive definite, then there exists a cartesian basis of (W, σ) whose vectors are mutually orthogonal with respect to τ.*

The proof proceeds as for Theorem 1 but here Proposition 3 makes Lemma 2 redundant.

THEOREM 3H.   *If f is an automorphism of the Hilbert space (W, σ), then there exists a cartesian basis all of whose vectors are eigenvectors of f.*

Again the proof is (this time, considerably) simpler than that of the euclidean result: our theorem follows directly from Proposition 3 and the complex analogue of Lemma 6.

The above definition of Hilbert space is one of the few in this book that does not generalize immediately to the infinite dimensional case. A basic property (automatically satisfied for finite dimensional Hilbert spaces) is that of *completeness*: if $\|a_m - a_n\| \to 0$ as $m, n \to \infty$, then there exists a vector a such that $\|a - a_n\| \to 0$ as $n \to \infty$. Infinite dimensional Hilbert spaces are of crucial importance in many branches of mathematics and particularly in functional analysis. See for example [2], [8].

EXERCISES

In the following exercises (W, σ) denotes a Hilbert space.

1. Rewrite the proofs of Lemmas 2 and 4 using $\sigma_{(C)}$ instead of σ'.

2. Prove that

$$4\sigma(a, b) = \|a+b\|^2 - \|a-b\|^2 + i\|a-ib\|^2 - i\|a+ib\|^2$$

for all a, b in W.

(If d is a mapping which satisfies the rules D.1–D.6 of § 6.1, p. 125 as modified on p. 146 of this section and if we write $\|a-b\| = d(a, b)$ then the above equation may be used to *define* a positive definite hermitian form σ (cf. exercises 4, 5 of § 6.1).)

3. Let τ be a sesquilinear form on W with respect to complex conjugation. Show that the equation

$$\sigma(af, b) = \tau(a, b) \quad \text{for all } a, b \text{ in } W$$

defines a linear mapping f of W into W. Deduce that there exists a cartesian basis of (W, σ) with respect to which τ has a triangular matrix.

4. If f, g are linear mappings of W into W such that $fg = gf$, show that there exists a cartesian basis of (W, σ) with respect to which the matrices of f and g are both triangular. (Cf. exercise 3 of § 6.3.)

5. Let f be a linear mapping of W into W. Show that the equation

$$\sigma(af^*, b) = \sigma(a, bf)$$

for a, b varying in W, defines a linear mapping f^* of W into W. Verify also that
    (i) $f^{**} = f$;
    (ii) $(f+g)^* = f^* + g^*$;
    (iii) $(fg)^* = g^*f^*$;
    (iv) $(xf)^* = \bar{x}f^*$;
    (v) $(f-x1)^* = f^* - \bar{x}1$.

6. Let the linear mapping f of W into W satisfy $ff^* = f^*f$ (in the notation of the previous exercise). We call such an f a *normal mapping* on W.

Prove

(i) $\|wf\| = \|wf^*\|$ for all w in W;

(ii) if w is an eigenvector of f with eigenvalue x, then w is an eigenvector of f^* with eigenvalue $\bar{x}$;

(iii) if $[w]f \subset [w]$ then $[w]^\perp f \subset [w]^\perp$.

Deduce the following

THEOREM. *If f is a normal mapping on the Hilbert space (W, σ) then there exists a cartesian basis all of whose vectors are eigenvectors of f.*

Observe that f is σ-symmetric if, and only if, $f^* = f$ and f is an automorphism of (W, σ) if, and only if, $ff^* = 1$. These mappings are often referred to in the literature as *hermitian* and *unitary* mappings, respectively, and are special cases of normal mappings. The reader should compare the above result with Proposition 4, p. 136 (from which we deduced Theorem 1) and Theorem 3H, p. 146.

CHAPTER VII

Modules

7.1 Rings and Modules

In Chapter IV we encountered the problem of finding the similarity classes of square matrices. An illuminating way of dealing with this problem is in terms of *modules*. These arise when we reconsider the notion of vector space by allowing the "scalars" to lie in a set having a more general algebraic structure than that of a field.

DEFINITION. Let R be a set together with rules of addition and multiplication that associate to any two elements x, y in R a sum $x+y$ and a product xy, both in R. Then R is called a *ring* if the following axioms are satisfied:

R.1. R is a commutative group with respect to addition;

R.2. $(xy)z = x(yz)$ for all x, y, z in R;

R.3. $x(y+z) = xy + xz$ and
$(x+y)z = xz + yz$,   for all x, y, z in R.

The ring R is *commutative* if

R.4. $xy = yx$ for all x, y in R.

A subset of R that is itself a ring (with respect to the same rules of addition and multiplication as R) is called a *subring* of R.

If R contains an element $1 \neq 0$ such that

$$x1 = 1x = x \quad \text{for all } x \text{ in } R,$$

then 1 is necessarily the only element with this property. We call 1 the *identity element* of R.

Obviously, every field F is a ring. So are $\mathbf{Z}$ and $F[X]$ and it is with these that we shall be primarily concerned. All our rings will therefore be commutative and possess an identity element. (Note, however, that $\mathscr{L}(V, V)$ is an example of a ring that is commutative if, and only if, $\dim V = 1$; and that the subring $m\mathbf{Z}$, of all integers divisible by m, has an identity if, and only if, $m = \pm 1$.)

149

The definition (given on p. 5) of a vector space may now be repeated word for word but using a commutative ring R with identity instead of the field F. We write this out for convenience.

DEFINITION. Let R be a given commutative ring with identity and M a set together with rules of addition and multiplication which associate to any two elements a, b in M a *sum* $a+b$ in M, and to any two elements x in R, a in M a *product* xa in M. Then M is called a *module over the ring R* or an *R-module* if the following axioms hold:

M.1. M is a commutative group with respect to addition;
M.2. $x(a+b) = xa+xb$,
M.3. $(x+y)a = xa+ya$,
M.4. $(xy)a = x(ya)$,
M.5. $1a = a$ where 1 is the identity element of R,
for all x, y in R and all a, b in M.

The following are examples of modules:

1. If R is a field, then an R-module is the same as a vector space over R.

2. If A is a commutative group with respect to addition, we may define na for any $n \neq 0$ in Z and a in A by the rule

$$na = \varepsilon(\underbrace{a+\cdots+a}_{|n|})$$

where $\varepsilon = \pm 1$ according as $n = \pm |n|$. If we also set $0a = 0$, then A is a Z-module. Hence *every commutative group may be regarded as a Z-module.*

3. Let R be a subring of the commutative ring S and let 1 be an identity element for both R and S. Then S is an R-module if we define the product of the module element a in S by x in R to be xa, their product as elements of the ring S. (Cf. example 2, p. 5.) In particular, R is a module over R.

4. Let R be a commutative ring with identity and denote by R^n the set of all n-tuples $(a_1, \ldots, a_n)$ where $a_i \in R$ for $i = 1, \ldots, n$. If we define addition and multiplication by "scalars" term by term (as for the vector space F^n on p. 5), then R^n is a module over R.

5. (The crucial example for the similarity problem.) Let V be a vector space over a field F and f a linear mapping of V into V. If

$$p(X) = c_0 X^n + c_1 X^{n-1} + \cdots + c_n$$

is an arbitrary element of $F[X]$ we may form

$$p(f) = c_0 f^n + c_1 f^{n-1} + \cdots + c_n 1_V,$$

which is an element of the linear algebra $\mathscr{L}(V, V)$. The following simple rules are satisfied:

$$\text{If} \quad p(X)+q(X) = r(X), \quad \text{then} \quad p(f)+q(f) = r(f); \tag{1}$$
$$\text{if} \quad p(X)q(X) = s(X), \quad \text{then} \quad p(f)q(f) = s(f). \tag{2}$$

The first rule is immediate and the second is true because the powers of f commute with the elements of F. These rules may be summarized by saying that the mapping $p(X) \to p(f)$ is a homomorphism of the linear algebra $F[X]$ into the linear algebra $\mathscr{L}(V, V)$.

We define a product $p(X)\, a$ for any a in V by the rule

$$p(X)\, a = a\, p(f).$$

Rules (1), (2) above now show that the additive group V with this multiplication is a module over $F[X]$. We shall write this module as V_f to distinguish it from the vector space V. (In checking axiom M.4 we need the commutativity of $F[X]$. We could equally well write $a\, p(X)$ for $p(X)\, a$, but prefer to maintain our convention of writing "scalars" on the left and linear mappings on the right.)

One essential way in which an R-module may differ from a vector space is as follows. If $xa = 0$ ($x \in R$, $a \in M$) and $x \neq 0$, then we may not be able to deduce that a is zero. For example $\mathbf{F}_p$ is a $\mathbf{Z}$-module in which $p1 = 0$.

For a less trivial illustration we return to the situation of example 5. Since $\mathscr{L}(V, V)$ has finite dimension over F, there is a smallest positive integer r such that f^r is linearly dependent on $f^{r-1}, f^{r-2}, \ldots, 1_V$: say

$$f^r = c_1 f^{r-1} + \cdots + c_r 1_V.$$

If we put

$$m(X) = X^r - c_1 X^{r-1} - \cdots - c_r$$

then $m(f) = 0$. Thus $m(X)\, a = a\, m(f) = 0$ for all a in V. Similarly, if a is a given non-zero vector in V, there is a smallest positive integer s such that af^s is linearly dependent on $af^{s-1}, \ldots, a$: say

$$af^s = d_1 af^{s-1} + \cdots + d_s a.$$

If we put

$$m_a(X) = X^s - d_1 X^{s-1} - \cdots - d_s$$

then $a\, m_a(f) = 0$. Thus $m_a(X)\, a = a\, m_a(f) = 0$. The polynomial $m(X)$ is called the *minimum polynomial* of f; and $m_a(X)$ is called the *minimum polynomial of f at a*.

DEFINITION. Let S be a subset of the R-module M. The set $\mathfrak{a}(S)$ of all x in R such that $xS = 0$ is called the *annihilator* of S.

If $S = \{a\}$, we write $\mathfrak{a}(a)$ for $\mathfrak{a}(S)$ and call this the annihilator of a. If $\mathfrak{a}(a) \neq 0$ we call a a *torsion element*. If every element in M is a torsion element we say M is a *torsion module*. If 0 is the only torsion element, M is said to be *torsion-free*.

Our remarks above have shown that $m(X)$ belongs to the annihilator $\mathfrak{a}(V_f)$ and that V_f is a torsion module. On the other hand, every vector space is torsion-free; so is the module $F[X]^n$, where F is a field (cf. example 4).

In order to examine annihilators in more detail we need a further notion.

DEFINITION. Let R be a commutative ring. A subgroup $\mathfrak{a}$ of the additive group R is called an *ideal* of R if $rx \in \mathfrak{a}$ whenever $r \in R$ and $x \in \mathfrak{a}$.

It follows at once from the definitions that the annihilator $\mathfrak{a}(S)$ is an ideal of the ring R. Note also that $\{0\}$ and R are ideals of R.

If x is any fixed element of R then the set Rx of all rx for r in R is clearly an ideal of R. Such an ideal is called *principal*. In particular R is a principal ideal provided R has an identity element.

PROPOSITION 1. *If R is a commutative ring with identity then R is a field if, and only if, $\{0\}$ and R are the only ideals of R.*

PROOF. Suppose that R is a field and that x is a non-zero element in an ideal $\mathfrak{a}$ of R. For any r in R the element $r\,x^{-1}x = r$ is in $\mathfrak{a}$ and so $\mathfrak{a} = R$.

Conversely, take any $x \neq 0$ in R and consider the principal ideal Rx. Since $x = 1x \in Rx$, $Rx \neq \{0\}$ and so by hypothesis $Rx = R$. Thus $yx = 1$ for some y in R, i.e., x has an inverse. It now follows easily that the non-zero elements in R form a group with respect to multiplication and hence R is a field.

PROPOSITION 2. *Every ideal of $\mathbf{Z}$ or $F[X]$ is principal.*

PROOF. Let $\mathfrak{a}$ be a non-zero ideal of $F[X]$ and p any non-zero element in $\mathfrak{a}$ of smallest degree. Any q in $\mathfrak{a}$ can be written in the form

$$q = up + v,$$

where degree $v <$ degree p. Since $up \in \mathfrak{a}$, $v \in \mathfrak{a}$ and so $v = 0$ by our choice of p. Hence $\mathfrak{a} = F[X]\,p$.

The proof for $\mathbf{Z}$ is similar but uses absolute values instead of degrees.

In view of the proof of Proposition 2 we see that the annihilator $\mathfrak{a}(V_f)$ is simply the set of all multiples of the minimum polynomial $m(X)$, i.e., the principal ideal $F[X]\,m(X)$; and also that $\mathfrak{a}(a) = F[X]\,m_a(X)$.

EXERCISES

1. Let V be the vector space F^2 and f the linear mapping of V into V whose matrix with respect to the standard basis is

$$\begin{pmatrix} 1 & 2 \\ 2 & 4 \end{pmatrix}.$$

If $a_1 = (1, 0)$, $a_2 = (0, 1)$, $a_3 = (2, -1)$, $a_4 = (1, 2)$ find $m_{a_i}(X)$ for $i = 1, 2, 3, 4$ and also $m(X)$.

2. Show that $m_a(X)$ divides $m(X)$ for all a in V. Hence show that if x is an eigenvalue of f, then $m(x) = 0$.

3. Show that all the subrings of $\mathbf{Z}$ are ideals of $\mathbf{Z}$.

4. If A is a commutative group and a is an element of finite order k, prove that $\mathbf{Z}k = \mathfrak{a}(a)$, where A is viewed as a $\mathbf{Z}$-module.
   If $\mathfrak{a}(A) = \mathbf{Z}t$ and A is finite of order s, prove that t divides s. (One calls t the *exponent* of A.)

7.2　Submodules and Homomorphisms

We define a submodule in exactly the same way as a subspace and now assert: *the whole of* § 1.3 (p. 6) *remains true for modules over a ring*.

A module that can be generated by a finite number of its elements is said to be *finitely generated*. If it can be generated by a single element it is called *cyclic*.

If we regard a commutative group A as a $\mathbf{Z}$-module then the submodules of A are just the subgroups of A. If R is a commutative ring with identity and R is regarded as an R-module (cf. example 3 above), then the submodules of R are the ideals of R. In particular, the cyclic submodule of R generated by the element x is the principal ideal Rx. In accordance with the notation introduced in § 1.3 we write $[x] = Rx$.

It is easy to characterize the submodules of V_f. If M is a submodule, then M must be a subspace of V (because $F \subset F[X]$) and $Mf \subset M$ (because $vf = Xv$). But conversely, any subspace M such that $Mf \subset M$ is clearly a submodule of V_f. The submodules of V_f are usually called the *invariant subspaces* of f or the f-invariant subspaces. Observe that an invariant subspace of dimension 1 is generated by an eigenvector of f. If S is a subset of V then we shall write the submodule of V_f generated by S as $[S]_f$ (to distinguish it from the subspace $[S]$ spanned by S).

PROPOSITION 3. *If $M = [a]$ is a cyclic $F[X]$-module whose annihilator $\mathfrak{a}(M)$ is generated by the non-zero polynomial $p(X)$ of degree s, then M is a vector space over F of dimension s.*

PROOF. The "vectors" a, Xa, ..., $X^{s-1}a$ are linearly independent over F (by the definition of $p(X)$) and the vectors $X^s a$, $X^{s+1}a$, ... are in turn seen to be linearly dependent on them. Thus a, Xa, ..., $X^{s-1}a$ form a basis of M over F.

In particular we see that $[a]_f$ is a subspace of V of dimension equal to the degree of $m_a(X)$.

For any commutative ring R with identity we may define a *homomorphism* or *R-homomorphism* of one R-module into another in exactly the same way as a linear mapping (cf. p. 66). Isomorphisms, automorphisms and the kernel of a homomorphism are all defined as before.

If K is a submodule of a module M we may introduce the cosets of K and turn the set of all cosets M/K into an R-module just as was done for vector spaces (p. 67). Proposition 1 of § 4.1 remains true for modules. The interesting new feature is that this construction can be carried out when $M = R$ and K is an ideal of R. Then R/K is a cyclic module, generated by $1 + K$, where 1 is the identity of R. We assert that every cyclic module arises in this way:

PROPOSITION 4.   *If M is a cyclic R-module, then M is isomorphic to $R/\mathfrak{a}(M)$.*

PROOF.   Let $M = [a]$ and consider the mapping

$$f: x \to xa$$

of R onto M. Clearly f is an R-homomorphism and the kernel is $\mathfrak{a}(a)$. Now $\mathfrak{a}(a) = \mathfrak{a}(M)$ and hence the result follows from the module version of Proposition 1 of § 4.1 (p. 67).

The reader will easily see that Proposition 3 above can be deduced from this result. It also follows that a torsion-free cyclic R-module is isomorphic to R itself.

DEFINITION.   If $f, g \in \mathscr{L}(V, V)$, we say f, g are *similar* if there exists a linear automorphism e of V such that $g = e^{-1}f e$.

Observe that f, g are similar if, and only if, their matrices (with respect to the same ordered basis) are similar (cf. p. 79).

PROPOSITION 5.   *If $f, g \in \mathscr{L}(V, V)$, then V_f, V_g are isomorphic if, and only if, f, g are similar.*

PROOF.   (i) If there is an $F[X]$-isomorphism e of V_f onto V_g, then e is a linear isomorphism of V onto V and also

$$(p(X) v) e = p(X) (ve)$$

for all v in V and all $p(X)$ in $F[X]$.   In other words,

$$(v\, p(f))e \,=\, (ve)p(g),$$

for all v in V and all $p(X)$ in $F[X]$.   Hence

$$p(g) \,=\, e^{-1}p(f)\, e$$

and in particular

$$g \,=\, e^{-1}f\, e.$$

(ii) If $g = e^{-1}f\, e$, where e is a linear automorphism of V, then

$$g^n \,=\, e^{-1}f^n\, e$$

for all $n \geq 0$ and so

$$p(g) \,=\, e^{-1}p(f)\, e.$$

Hence, by reversing the argument in (i), we see that e is an isomorphism of V_f onto V_g.

EXERCISES

1. Let R be a commutative ring with identity and suppose that $xy = 0$ always implies that at least one of x and y is 0.   (We call such a ring an *integral domain*: cf. also the end of § 7.3.)

   Prove that, if M is a finitely generated torsion module over R, then $\mathfrak{a}(M) \neq 0$.

2. Prove that $M = \mathbf{Q}/\mathbf{Z}$ is a torsion module over $\mathbf{Z}$ but that $\mathfrak{a}(M) = 0$.
   (This shows that finite generation is necessary in exercise 1.)

3. If R is an integral domain (see exercise 1) and M is a module over R, prove that the set of all torsion elements in M forms a submodule.

   This result is false without the integral domain property: if $R = \mathbf{Z}/[6]$, show that $(2, 0)$, $(0, 3)$ are torsion elements in R^2 but that their sum is not a torsion element.

7.3 Direct Decompositions

Our aim is to determine completely the structure of finitely generated $F[X]$-modules.

THEOREM 1.   *Let F be a field and M a finitely generated $F[X]$-module. Then M can be decomposed into a direct sum of cyclic submodules*

$$M \,=\, [a_1] \oplus \cdots \oplus [a_k],$$

where $\mathfrak{a}(a_1) \subset \mathfrak{a}(a_2) \subset \cdots \subset \mathfrak{a}(a_k)$ and $\mathfrak{a}(a_k) \neq F[X]$.
The ideals $\mathfrak{a}(a_i)$, $i = 1, \ldots, k$, are uniquely determined by M.

We remark that if M is the zero module then the theorem is trivially true (with $k = 0$).

DEFINITION. The ideals $\mathfrak{a}(a_i)$, $i = 1, \ldots, k$, are called the *invariant factors* of M. The monic generators of these ideals are also frequently called the invariant factors of M.

PROOF. We note that if $b_1, \ldots, b_l$ generate M, then so do

$$b_1, \ldots, b_{l-1}, b_l + q_j b_j,$$

where $q_j \in F[X]$ and $1 \le j \le l - 1$. Let k be the smallest number of elements needed to generate M.

(a) Suppose first that we can find a set $\{b_1, \ldots, b_k\}$ of generators among which no non-trivial relation holds: i.e., such that

$$m_1 b_1 + \cdots + m_k b_k = 0$$

implies $m_1 = \cdots = m_k = 0$. Then

$$M = [b_1] \oplus \cdots \oplus [b_k]$$

and $\mathfrak{a}(b_i) = 0$ for each i. The first part of the theorem is now established.

(b) Assume next that every system of k generators gives at least one non-trivial relation. Then there is one such system $\{b_1, \ldots, b_k\}$ for which a relation

$$m_1 b_1 + \cdots + m_k b_k = 0$$

holds with a non-zero coefficient m_i of minimum degree. We may suppose that $i = k$.

We prove first that

$$m_j \text{ is a multiple of } m_k \text{ for } j = 1, \ldots, k-1. \tag{i}$$

For if

$$m_j = q_j m_k + r_j$$

where degree $r_j <$ degree m_k, then

$$m_1 b_1 + \cdots + r_j b_j + \cdots + m_k(b_k + q_j b_j) = 0,$$

from which we deduce that $r_j = 0$.

We contend also that

$$\text{if } n_1 b_1 + \cdots + n_k b_k = 0 \text{ then } n_k \text{ is a multiple of } m_k.$$

For if

$$n_k = q\, m_k + r \tag{ii}$$

where degree $r <$ degree m_k, then by subtracting q times the first relation from the second we obtain

$$(n_1 - q\, m_1) b_1 + \cdots + r\, b_k = 0,$$

whence $r = 0$.

If we define

$$a_k = q_1 b_1 + \cdots + q_{k-1} b_{k-1} + b_k,$$

then (i) shows that

$$m_k a_k = 0. \tag{iii}$$

If M' denotes the submodule of M generated by $b_1, \ldots, b_{k-1}$ then

$$M = M' \oplus [a_k]. \tag{iv}$$

This follows since the relation

$$n_1 b_1 + \cdots + n_{k-1} b_{k-1} + n_k a_k = 0$$

implies that n_k is a multiple of m_k (by (ii)) and so $n_k a_k = 0$ (by (iii)).

The first part of our theorem may now be proved by induction on k. We can express M' as a direct sum

$$[a_1] \oplus \cdots \oplus [a_{k-1}]$$

where $\mathfrak{a}(a_1) \subset \cdots \subset \mathfrak{a}(a_{k-1})$.

It only remains to prove that

$$\mathfrak{a}(a_{k-1}) \subset \mathfrak{a}(a_k). \tag{v}$$

We proved under (i) that if we have a relation between k generators involving m_k as a coefficient, then all coefficients are divisible by m_k. If now $x \in \mathfrak{a}(a_{k-1})$ then $x a_{k-1} = 0$ and so $x a_{k-1} + m_k a_k = 0$. Hence m_k divides x.

In order to prove the uniqueness part of the theorem we suppose that r of the ideals $\mathfrak{a}(a_i)$ are zero $(0 \le r \le k)$ and define

$$M_1 = [a_1] \oplus \cdots \oplus [a_r],$$
$$M_2 = [a_{r+1}] \oplus \cdots \oplus [a_k].$$

We assert that M_2 consists of all the torsion elements of M and thus is uniquely determined by M. For if $a = p_1 a_1 + \cdots + p_k a_k$ and if $q a = 0$ where q is a non-zero element of $F[X]$, then it follows from the direct decomposition that $q p_1 = \cdots = q p_r = 0$ and hence $p_1 = \cdots = p_r = 0$. Therefore $a \in M_2$.

Using the decomposition $M = M_1 \oplus M_2$ we can write any a in M in the form $a = a_1 + a_2$ where $a_1 \in M_1$ and $a_2 \in M_2$. The mapping $a \to a_1$ is a homomorphism of M onto M_1 with kernel M_2 and so M/M_2 is isomorphic to M_1. Thus M_1 is determined by M to within isomorphism. (Note that we have *not* shown that the submodule M_1 is uniquely determined by M: cf. exercise 9 below.)

The proof of the uniqueness of the invariant factors of M is now reduced to the special cases where M is either a torsion module or is torsion-free. The former case is postponed till we prove Theorem 2. Let us therefore assume that

$$M = [a_1] \oplus \cdots \oplus [a_k]$$

where $a(a_i) = 0$ for each i. Our aim is to show that the integer k is uniquely determined by M.

Distinct elements $b_1, \ldots, b_r$ of M are called *linearly independent over* $F[X]$ if every relation

$$m_1 b_1 + \cdots + m_r b_r = 0 \quad (m_i \in F[X])$$

implies that each m_i is zero. The mapping

$$p_1 a_1 + \cdots + p_k a_k \;\rightarrow\; (p_1, \ldots, p_k)$$

is an $F[X]$-isomorphism of M onto $F[X]^k$ and so preserves linear independence over $F[X]$. Let $F(X)$ denote the field of quotients $f(X)/g(X)$ where $f(X)$, $g(X) \in F[X]$ and $g(X) \neq 0$. Then the linear independence of elements of $F[X]^k$ is equivalent to their linear independence as elements of the containing vector space $F(X)^k$ (for any common denominators can be multiplied out). It follows that k is the maximum number of linearly independent elements of M over $F[X]$ and as such is an invariant of M.

PROPOSITION 6. *Let M be a cyclic $F[X]$-module whose annihilator $a(M)$ is generated by the non-zero polynomial p. If $p_1, \ldots, p_n$ are the distinct monic irreducible factors of p, then*

$$M = [a_1] \oplus \cdots \oplus [a_n],$$

where $a(a_i)$ is generated by a power of p_i $(i = 1, \ldots, n)$.

Let $M = [a]$ and suppose that $p = s\,t$, where s, t are relatively prime. If $b = t\,a$, then $a(b) = [s]$: for $q\,b = 0$ implies $q\,t \in a(a)$ and so p divides $q\,t$, i.e., s divides q. Similarly, if $c = s\,a$, $a(c) = [t]$. Since s and t are relatively prime we have

$$us + vt = 1$$

for some u, v in $F[X]$ (cf. exercise 3 below). This implies immediately that $[a] = [b] + [c]$. Further, since any element of $[b] \cap [c]$ is annihilated by s and t, we must have $[b] \cap [c] = 0$ and so $[a] = [b] \oplus [c]$. The result now follows by induction on n (the number of distinct irreducible factors of p).

THEOREM 2. *Let F be a field and M a finitely generated $F[X]$-module whose annihilator $a(M)$ is generated by the non-zero polynomial m. If*

$p_1, \ldots, p_t$ are the distinct monic irreducible factors of m, then M may be expressed as the direct sum

$$M = K_1 \oplus \cdots \oplus K_t$$

where K_i is the submodule of all elements in M annihilated by some power of p_i.

Each submodule K_i may be expressed as a direct sum

$$K_i = [a_{i1}] \oplus \cdots \oplus [a_{ik_i}]$$

where $\mathfrak{a}(a_{i1}) \subset \mathfrak{a}(a_{i2}) \subset \cdots \subset \mathfrak{a}(a_{ik_i})$.

The ideals $\mathfrak{a}(a_{ij})$ $(i = 1, \ldots, t; j = 1, \ldots, k_i)$ are uniquely determined by M.

DEFINITION. The ideals $\mathfrak{a}(a_{ij})$ (or frequently also their monic generators) are called the *elementary divisors* of M.

We note as a consequence of the uniqueness of the elementary divisors that the submodule $[a_{ij}]$ cannot be decomposed further into the direct sum of proper submodules. Such a submodule is called *indecomposable*. The decomposition of M corresponding to the elementary divisors is into as many cyclic summands as possible while the decomposition corresponding to the invariant factors is into as few as possible.

PROOF. By applying first Theorem 1 and then Proposition 6 we may decompose M into a direct sum of cyclic submodules each of which is annihilated by some power of an irreducible factor of m. The set K_i of all "vectors" annihilated by some power of p_i is therefore the direct sum of those cyclic submodules that are annihilated by some power of p_i. We may number these submodules

$$[a_{i1}], \ldots, [a_{ik_i}],$$

where

$$\mathfrak{a}(a_{ij}) = [p_i{}^{n_{ij}}] \quad \text{and} \quad n_{i1} \geq n_{i2} \geq \cdots \geq n_{ik_i}.$$

Since the submodules K_i are uniquely determined by M, it is sufficient in proving the uniqueness of the elementary divisors to consider the case where m is a power of an irreducible polynomial, say of p. Then

$$M = [a_1] \oplus \cdots \oplus [a_k]$$

where

$$\mathfrak{a}(a_j) = [p^{n_j}] \quad \text{and} \quad n_1 \geq n_2 \geq \cdots \geq n_k.$$

Thus $\mathfrak{a}(M) = [p^{n_1}]$.

We shall now argue by induction on n_1. The mapping $a \to pa$ is an $F[X]$-homomorphism of M into M. Let N be the kernel. Then

$$\sum_{j=1}^{k} q_j a_j \in N$$

if, and only if, q_j is a multiple of $p^{n_j - 1}$, for $j = 1, \ldots, k$. Hence N is the direct sum of k cyclic submodules each of which has annihilator $[p]$. By Proposition 3 we see that

$$\dim_F N = k \ (\text{degree } p)$$

and so k is uniquely determined by M. Now $p^{n_1 - 1}$ generates $\mathfrak{a}(pM)$ and so by the induction hypothesis, the integers

$$n_1 - 1 \geq n_2 - 1 \geq \cdots \geq n_{k'} - 1 > 0$$

are determined uniquely. Finally, the sequence

$$n_1 \geq n_2 \geq \cdots \geq n_{k'} > n_{k'+1} = \cdots = n_k = 1$$

is uniquely determined by M.

The last part of Theorem 1 now follows since the invariant factor $\mathfrak{a}(a_j)$ of Theorem 1 is generated by $\prod_i p_i^{n_{ij}}$, where the product is taken over all i for which n_{ij} is defined.

PROPOSITION 7. *Two finitely generated modules over $F[X]$ are isomorphic if, and only if, they have the same invariant factors.*

Two finitely generated torsion modules over $F[X]$ are isomorphic if, and only if, they have the same elementary divisors.

This result is an immediate consequence of Theorems 1 and 2 together with Proposition 4.

Theorems 1 and 2 are also true for modules over $\mathbf{Z}$ and thus give the complete structure of finitely generated commutative groups. Our proofs go over word for word provided only that we use the absolute value of an integer instead of the degree of a polynomial and replace monic irreducible polynomials by positive prime numbers.

The results of this section may be generalized as follows. Let R be a commutative ring with identity in which $xy = 0$ implies either $x = 0$ or $y = 0$ (or both). Such a ring is called an *integral domain*. The construction leading from the ring of integers $\mathbf{Z}$ to the field of rationals $\mathbf{Q}$ may be applied to R to yield a field, called the *field of quotients* of R. (For example, $F(X)$ is the field of quotients of $F[X]$.) An integral domain R in which every ideal is principal is called a *principal ideal domain*. We shall see in exercises 1 to 4 below that such a principal ideal domain is a *unique factorization domain*, i.e., any non-invertible element may be expressed as a product of irreducible factors and this factorization is essentially unique. This fact allows us to replace $F[X]$ by R in the statement of Theorem 2 (where $p_1, \ldots, p_t$ are only determined to within invertible elements). *Then Theorems 1 and 2 are true for finitely genera-ted modules over an arbitrary principal ideal domain.* (Cf. exercises 7, 8.)

EXERCISES

The ring R in Exercises 1–8 is assumed to be a principal ideal domain.

1. Given an ascending chain of ideals of R

$$a_1 \subset a_2 \subset \cdots$$

show that their union $\bigcup a_i$ is an ideal of R. Deduce that the chain terminates, i.e., there is an integer n such that

$$a_n = a_{n+r} \quad \text{for all } r \geq 0.$$

2. Given two non-zero elements y, z in R, we say that y *divides* z if there exists in R an element x such that $z = xy$, i.e., if $Rz \subset Ry$. If y divides 1 then y is said to be *invertible*. An element z is called *irreducible* if it is not invertible and if $z = xy$ implies that x or y is invertible.

   Prove that $Rz = Ry$ if, and only if, z and y are related by an equation $z = xy$ where x is invertible. Using this and exercise 1, show that any non-zero non-invertible element of R can be expressed as a product of a finite number of irreducible elements.

3. Given s, t in R, show that $[d] = [s] + [t]$ if, and only if, d satisfies the following two conditions:
   (i) d divides s and t;
   (ii) any element which divides s and t must also divide d.

   An element d satisfying (i) and (ii) is called a *greatest common divisor* of s and t. If d' is another greatest common divisor of s and t, then $d' = ed$ where e is invertible. Moreover, $d = us + vt$ for suitable u, v in R.

4. Use exercise 3 to show that, if $p_1 \ldots p_r = q_1 \ldots q_s$, where each p_i and each q_i is irreducible, then $r = s$ and, with a suitable renumbering of $q_1, \ldots, q_s$, $p_i = e_i q_i$, where each e_i is invertible.

   (Let R be any integral domain. We define invertible and irreducible elements as was done in exercise 2. If R is such that every non-zero, non-invertible element can be written as the product of a finite number of irreducible elements and if this expression is unique to within order and multiplication by invertible elements, then R is called a *unique factorization domain*. Exercises 2 and 4 together show that every principal ideal domain is a unique factorization domain.)

5. We define the *length* of a non-zero element x in R as follows: if x is inver-tible, x has length 0; if $x = p_1 \ldots p_r$ where each p_i is irreducible, then x has length r. (This definition is unambiguous by exercise 4 and applies to every non-zero element of R by exercise 2.)

   Show that if y divides z and the lengths of y and z are equal, then z also divides y.

6. Let $d \neq 0$ be a greatest common divisor of s and t and suppose $d = us + vt$ (cf. exercise 3). Show that the inverse of the matrix

$$\begin{pmatrix} v & -u \\ s/d & t/d \end{pmatrix} \quad \text{is} \quad \begin{pmatrix} t/d & u \\ -s/d & v \end{pmatrix}.$$

If an R-module M is generated by elements a and b where the relation $sa + tb = 0$ is satisfied, prove that M is also generated by $a' = va - ub$ and $b' = (s/d)a + (t/d)b$ where the relation $db' = 0$ is satisfied.

7. In part (b) of the proof of Theorem 1 assume that

$$m_1 b_1 + \cdots + m_k b_k = 0$$

is a relation in which m_k is a non-zero coefficient of *minimum length* (as defined in exercise 5 above). Use exercises 5 and 6 to adapt the proof of Theorem 1 to the case of an arbitrary principal ideal domain.

8. State and prove the generalization of Theorem 2 to a finitely generated module over a principal ideal domain.

9. Let B, C be commutative groups with respect to addition and let A be the set of all ordered pairs (b, c) where $b \in B$, $c \in C$. Define

$$(b, c) + (b', c') = (b + b', c + c')$$

and show that A is a commutative group with respect to this addition. If $B = \mathbf{Z}$, $C = \mathbf{F}_2$ and $a_1 = (1, 0)$, $a_2 = (0, 1)$, $a_1' = (1, 1)$, show that

$$A = [a_1] \oplus [a_2] \quad \text{and} \quad A = [a_1'] \oplus [a_2].$$

(This gives an example where the direct decomposition of Theorem 1 is not unique.)

10. If F is a finite field, deduce from Theorem 2 that the additive group of F is of the form $\mathbf{F}_p \oplus \cdots \oplus \mathbf{F}_p$ for some prime number p. Hence F has p^k elements for some k. (Cf. exercise 8 of § 5.4.)

11. Let F^* denote the multiplicative group of the finite field F with q elements. Show that there is an integer m such that
   (i) $x^m = 1$ for all x in F^*,
   (ii) there is an element of F^* of order m.
   Deduce that F^* is cyclic and that every element of F is a root of the polynomial $X^q - X$.

12. (i) If F is a field containing $\mathbf{F}_p$ and E is the subset of all elements x that satisfy $x^{p^m} = x$, then E is a field and contains $\mathbf{F}_p$.
   (ii) If the field F has p^k elements and m is a positive divisor of k, prove that there is one and only one field E with p^m elements contained in F. (Use (i) and exercise 11.)
   (iii) With F as in (ii), show that any field contained in F has p^m elements where m divides k.

   (We mention also that for every prime p and every positive integer k there does exist a field with p^k elements and any two such fields are isomorphic. These facts (which are not difficult to prove: see, e.g., the last chapter of G. Birkhoff and S. MacLane: *A Survey of Modern Algebra* (Macmillan)) together with the results of exercises 10, 11 and 12 constitute the whole elementary theory of finite fields.)

13. If G is a finite group of order $p_1{}^{n_1} \ldots p_t{}^{n_t}$ where $p_1, \ldots, p_t$ are distinct prime numbers, then a subgroup of G of order $p_i{}^{n_i}$ is called a *Sylow p_i-subgroup* of G. Show that a finite *commutative* group has exactly one Sylow p-subgroup for each prime p dividing its order, and that the group is the direct sum of its Sylow subgroups.

14. Let f belong to $\mathscr{L}(V, V)$ and let $m(X)$ be the minimum polynomial of f. If c is a non-zero scalar, show that the minimum polynomial of cf is $c^r m(X/c)$, where $r = \text{degree } m(X)$.

    If the elementary divisors of f are $(X - x_i)^{n_{ij}}$, $i = 1, \ldots, t$; $j = 1, \ldots, k_i$, prove that the elementary divisors of cf are $(X - cx_i)^{n_{ij}}$ with the same range of i and of j.

7.4   Equivalence of Matrices over $F[X]$

Theorem 1 shows that finitely generated *torsion-free* $F[X]$-modules have many of the properties of vector spaces. If E is such a module, then E can be written as a direct sum of, say, n cyclic submodules, each isomorphic to $F[X]$. The number n is an invariant of E, called the *rank* of E. It plays a role very similar to the dimension in vector space theory. Every minimal set of generators $\{v_1, \ldots, v_n\}$ is a maximal linearly independent set (but *not* conversely!) and $E = [v_1] \oplus \cdots \oplus [v_n]$. For obvious reasons we shall call $\{v_1, \ldots, v_n\}$ a *basis* of E. Given any $F[X]$-module M and any mapping g of $\{v_1, \ldots, v_n\}$ into M, then g can be uniquely extended to an $F[X]$-homomorphism of E into M, viz.

$$p_1 v_1 + \cdots + p_n v_n \to p_1(v_1 g) + \cdots + p_n(v_n g).$$

We may now assert that the whole of § 4.3 applies unchanged to finitely generated torsion-free $F[X]$-modules. So does the first part of § 4.4, including Proposition 4. But Theorem 2 fails and it is now our intention to find a substitute for this in the present theory.

    Let g be a homomorphism of the torsion free $F[X]$-module E of rank m into the torsion-free $F[X]$-module E' of rank n. Suppose there exist ordered bases (e_i), (e_j') of E, E' so that

$$(g; (e_i), (e_j')) = \begin{pmatrix} D & 0 \\ 0 & 0 \end{pmatrix},$$

where D is the diagonal matrix

$$\begin{pmatrix} d_1(X) & & 0 \\ & \ddots & \\ 0 & & d_t(X) \end{pmatrix}$$

and (i) d_i divides d_{i+1}, $i = 1, \ldots, t-1$, (ii) $d_t \neq 0$. Thus $d_1 e_1', \ldots, d_t e_t'$ generate the image Eg. If $d_1, \ldots, d_s$ are (non-zero) constants but d_{s+1} has degree ≥ 1, then

$$E'/Eg = [e_{s+1}' + Eg] \oplus \cdots \oplus [e_n' + Eg]$$

and

$$[d_{s+1}] \supset \cdots \supset [d_t] \supset \underbrace{0 = \cdots = 0}_{n-t}$$

are the invariant factors of E'/Eg. But these are independent of (e_i), (e_j'). We conclude that all matrices of the above form that represent g must give the same chain of ideals in $F[X]$; and so the elements $d_i(X)$ are determined by g to within non-zero constant multiples.

Now we ask: does g always have a matrix of this form and if so how can it be calculated? The answer is essentially contained in our proof of Theorem 1.

Let (v_i), (v_j') be any ordered bases of E, E', respectively, and suppose

$$(g; (v_i), (v_j')) = C.$$

We restrict attention to the ordered bases of E that can be constructed from (v_i) by moves of the following two types:

(i) an interchange of two generators;

(ii) the addition of a multiple of one generator to another.

Let us call a move of either type an *elementary move*. Similarly, we only allow ordered bases of E' that are obtainable from (v_j') by elementary moves. Observe that an elementary move on (v_i) changes C by either an interchange of two rows or the addition of a multiple of one row to another. An elementary move on (v_j') changes the columns of C in a similar manner.

Now $b_j = v_j' + Eg$, $j = 1, \ldots, n$, generate E'/Eg and the rows of C are the coefficients in m relations satisfied by $b_1, \ldots, b_n$. This is the link with Theorem 1.

We may obviously assume that $g \neq 0$. Among all the allowed ordered bases of E and E' choose a pair (u_i), (u_j') so that the corresponding matrix of g has a non-zero term, say d_1, of minimum degree. Without loss of generality, d_1 can be taken in the $(1, 1)$ position. Then the argument under (b) in the proof of Theorem 1 shows that every term in the first row and first column is divisible by d_1 and hence, by further elementary moves, we obtain a matrix in which d_1 is the only non-zero term in the first row and first column. An induction on n now yields a matrix for g of the required form.

Summing up, we have proved the following result.

PROPOSITION 8. *Let (v_i), (v_j') be ordered bases of the torsion-free $F[X]$-modules E, E' of ranks m, n, respectively. If g is a homomorphism of E into E', then there exist ordered bases (e_i), (e_j') obtained from (v_i), (v_j'), respectively, by elementary moves and such that*

$$(g; (e_i), (e_j')) = \begin{pmatrix} D & 0 \\ 0 & 0 \end{pmatrix}$$

where D is the diagonal matrix

$$\begin{pmatrix} d_1 & & 0 \\ & \ddots & \\ 0 & & d_t \end{pmatrix}$$

and

(i) *d_i divides d_{i+1}, $i = 1, \ldots, t-1$,*
(ii) *$d_t \neq 0$.*

Moreover, if s is the largest index i for which $[d_i] = F[X]$ then

$$[d_{s+1}] \supset \cdots \supset [d_t] \supset \underbrace{0 = \cdots = 0}_{n-t}$$

are the invariant factors of E'/Eg.

Proposition 8 solves the equivalence problem for matrices with coefficients in $F[X]$ (cf. § 4.4). In the next section we shall use this to solve the similarity problem for matrices over F.

Proposition 8 remains true when $F[X]$ is replaced by $\mathbf{Z}$: cf. the penultimate paragraph of § 7.3.

EXERCISES

1. Let $(c_1, \ldots, c_l)$ be an ordered set of generators of an $F[X]$-module M. Prove that there exists an ordered set of generators $(a_1, \ldots, a_l)$, obtainable from $(c_1, \ldots, c_l)$ by elementary moves (i.e., moves that consist of either an interchange of two generators or the addition of a multiple of one generator to another) and such that

$$M = [a_1] \oplus \cdots \oplus [a_l],$$

where $\mathfrak{a}(a_1) \subset \cdots \subset \mathfrak{a}(a_l)$.

(Reexamine the decomposition part of the proof of Theorem 1. If $l > k$, we obtain $\mathfrak{a}(a_{k+1}) = \cdots = \mathfrak{a}(a_l) = F[X]$ and thus $a_{k+1} = \cdots = a_l = 0$.)

2. Let E be a torsion-free $F[X]$-module of rank n and S a submodule. Given an ordered basis (v_i) of E, prove (by using exercise 1) that there exists an ordered basis (e_i), obtainable from (v_i) by elementary moves, such that

$$S = [d_1 e_1] \oplus \cdots \oplus [d_n e_n],$$

where $d_1, \ldots, d_n$ lie in $F[X]$ and d_{i+1} divides d_i for $i = 1, \ldots, n-1$. Thus S is finitely generated and its rank is at most n.

3. Let E and S be as in exercise 2. Prove that
   (i) S has rank n if, and only if, E/S is a torsion module;
   (ii) there exists a submodule T so that $E = S \oplus T$ if, and only if, E/S is torsion-free.

4. If S is the subgroup of $\mathbf{Z}^4$ generated by the elements $(11, -10, -4, -7)$, $(3, -4, 2, 5)$, $(3, -4, -4, -7)$ show that $\mathbf{Z}^4/S$ is isomorphic to

$$\mathbf{Z} \oplus \mathbf{Z}/[6] \oplus \mathbf{Z}/[2].$$

7.5 Similarity of Matrices over F

Let f be a linear mapping of the vector space V into V. We shall use the direct decompositions of V_f given by Theorems 1 and 2 to find bases of V for which the matrices of f are as simple as possible.

The invariant factors and the elementary divisors of V_f are usually called the *invariant factors* and *elementary divisors* of f.

LEMMA 1. *If V_f is a cyclic module with generator a, say, then V has an ordered basis with respect to which f has matrix*

$$B = \begin{pmatrix} 0 & 1 & & & \\ & 0 & 1 & & \\ & & \ddots & & \\ & & & 0 & 1 \\ d_s & \cdots & & d_2 & d_1 \end{pmatrix},$$

where $m_a(X) = X^s - d_1 X^{s-1} - \cdots - d_s$.

PROOF. Take the ordered basis $(a, af, \ldots, af^{s-1})$ of Proposition 3.

THEOREM 3. *V has an ordered basis with respect to which f has matrix*

$$\begin{pmatrix} B_1 & & 0 \\ & \ddots & \\ 0 & & B_k \end{pmatrix}$$

where each B_i is of the form B of Lemma 1.

The matrix of Theorem 3 is often called a *rational canonical matrix* of f. The result is a consequence of the decomposition of Theorem 1.

If we suppose that the minimum polynomial $m(X)$ can be factorized in $F[X]$ into *linear* factors, then a simpler matrix can be chosen for f. (This will be true—for all f—when $F = \mathbf{C}$.)

LEMMA 2. *If V_f is a cyclic module $[a]_f$, where $m_a(X) = (X - x)^n$, then there is an ordered basis of V for which f has matrix*

$$C = \begin{pmatrix} x & 1 & & & 0 \\ & x & 1 & & \\ & & \ddots & & \\ & & & x & 1 \\ 0 & & & & x \end{pmatrix}.$$

PROOF. It follows from the definition of the polynomial $m_a(X)$ that $a, (X-x)a, \ldots, (X-x)^{n-1}a$ are linearly independent. Writing now $v_i = (X-x)^{i-1}a$, $i = 1, \ldots, n$, we have

$$(X-x)v_i = v_{i+1}, \quad i = 1, \ldots, n-1,$$

and

$$(X-x)v_n = 0.$$

Thus

$$v_1 f = xv_1 + v_2$$
$$\cdots \qquad\qquad \cdots$$
$$v_{n-1}f = \qquad\qquad xv_{n-1} + v_n$$
$$v_n f = \qquad\qquad\qquad xv_n,$$

i.e.,

$$(f; (v_i), (v_j)) = C.$$

Theorem 2 and Lemma 2 now give immediately

THEOREM 4. *If the minimum polynomial of f is*

$$(X-x_1)^{s_1} \ldots (X-x_t)^{s_t}$$

then an ordered basis of V may be chosen with respect to which f has matrix

$$\begin{pmatrix} A_1 & & 0 \\ & \ddots & \\ 0 & & A_t \end{pmatrix}$$

where, for each i, A_i is a square matrix

$$\begin{pmatrix} C_{i1} & & 0 \\ & \ddots & \\ 0 & & C_{ik_i} \end{pmatrix}$$

and, for each j, C_{ij} is the $n_{ij} \times n_{ij}$ matrix

$$\begin{pmatrix} x_i & 1 & & & 0 \\ & x_i & 1 & & \\ & & \ddots & & \\ & & & x_i & 1 \\ 0 & & & & x_i \end{pmatrix}.$$

The matrix of f given in Theorem 4 is called a *Jordan* (or *classical*) *canonical matrix* of f. Note that we have used the notation of Theorem 2: thus the elementary divisors of f are

$$(X-x_i)^{n_{ij}}, \ i=1, \ldots, t; \ j=1, \ldots, k_i.$$

Let $A = (a_{ij})$ be an arbitrary matrix in $F^{n \times n}$. Then for any vector space V of dimension n over F and any ordered basis $(v_1, \ldots, v_n)$ of V, A determines (as usual) the linear mapping f by the equations

$$v_i f = \sum_{j=1}^{n} a_{ij} v_j, \quad i = 1, \ldots, n.$$

Any two modules V_f obtained in this way from A are isomorphic (Proposition 5, p. 154). We may therefore define the *minimum polynomial*, the *invariant factors* and the *elementary divisors of the matrix A relative to F* to be those of f. (Cf. exercise 3.)

It follows from Proposition 7 that two square matrices have the same invariant factors (or the same elementary divisors) if, and only if, they are similar. In the special case $F = \mathbf{C}$, every square matrix is similar to a Jordan canonical matrix.

There remains the problem of how to calculate the invariant factors (or elementary divisors) of a given matrix or linear mapping.

Suppose $(v_1, \ldots, v_n)$ is an ordered basis of V and $(f; (v_i), (v_j)) = (a_{ij})$. Then

$$(p_1(X), \ldots, p_n(X)) \rightarrow v_1 p_1(f) + \cdots + v_n p_n(f)_{\bullet}$$

is a module homomorphism of $F[X]^n$ onto V_f with kernel S, say. Write $E = F[X]^n$ and let e_i be the n-tuple all of whose terms are 0 except the i-term, which is 1. Then $(e_1, \ldots, e_n)$ is the "standard basis" of E. If

$$y_i = X e_i - \sum_{j=1}^{n} a_{ij} e_j, \quad i = 1, \ldots, n,$$

and S' is the submodule generated by $y_1, \ldots, y_n$, then clearly $S' \subset S$. Moreover, the subspace W of E/S' spanned by $e_1 + S', \ldots, e_n + S'$ satisfies $XW \subset W$ and hence is a submodule. But $e_1 + S', \ldots, e_n + S'$ generate E/S' as $F[X]$-module and so $W = E/S'$. Thus every element e of E can be written in the form

$$e = \sum x_i e_i + y,$$

where $y \in S'$ and $x_1, \ldots, x_n \in F$. The image of e in V is $\sum x_i v_i$, which is zero if, and only if, $x_1 = \cdots = x_n = 0$. Hence $S' = S$. We now have

PROPOSITION 9. *If $(f; (v_i), (v_j)) = A$ and S is the submodule of $F[X]^n$ generated by the rows of the matrix $X I_n - A$, then the modules $F[X]^n/S$ and V_f are isomorphic.*

This result, together with Proposition 8 gives an explicit method of calculating the invariant factors of f (or A).

We conclude this section with a result needed at the end of § 7.6. If $f \in \mathcal{L}(V, V)$, we define a linear mapping f^t of V^* into V^* (the *transpose* of f) by the equation

$$v(\varphi f^t) = (vf)\varphi,$$

where $\varphi \in V^*$ and $v \in V$. (This is a special case of a situation studied in exercise 7, § 4.6, p. 83.)

PROPOSITION 10.   *V_f is isomorphic to $(V^*)_{f^t}$.*

PROOF. Let $V_f = [a_1]_f \oplus \cdots \oplus [a_k]_f$ be the decomposition according to Theorem 1 and write $M_i = [a_i]_f$. If

$$N_i = \cap(M_j^{\circ} : j \neq i),$$

then

$$(m_1 + \cdots + m_k)(v_1 + \cdots + v_k) = m_1 v_1 + \cdots + m_k v_k$$

(where $m_i \in M_i$, $v_i \in N_i$) and therefore the subspace of V^* spanned by the N_i's is their direct sum. But dim $N_i =$ dim M_i for all i and thus $V^* = N_1 \oplus \cdots \oplus N_k$. By the definition of f^t and because all M_i are f-invariant, all N_i are f^t-invariant. Thus

$$(V^*)_{f^t} = (N_1)_{f^t} \oplus \cdots \oplus (N_k)_{f^t}$$

and we show that this is the invariant factor decomposition of $(V^*)_{f^t}$.

If $v \in N_i$, then $v = 0$ if, and only if, $M_i v = 0$. Hence $p \in \mathfrak{a}(N_i)$ if, and only if, $M_i(pv) = 0$ for all v in N_i, i.e., if, and only if, $(pM_i)v = 0$ for all v in N_i, i.e., if, and only if, $pM_i = 0$. Thus $\mathfrak{a}(N_i) = \mathfrak{a}(M_i)$. Finally, since M_i is a cyclic $F[X]$-module and dim $M_i =$ dim N_i, it follows that N_i is also cyclic (Proposition 3).

Thus we see that V_f and $(V^*)_{f^t}$ have the same invariant factors and hence, by Proposition 7, they are isomorphic.

EXERCISES

1. Let $(v_1, \ldots, v_n)$ be an ordered basis of the vector space V over $\mathbb{C}$ and f be the linear mapping that takes v_1 to v_2, $\ldots v_{n-1}$ to v_n and v_n to v_1. Find a rational canonical matrix and a Jordan canonical matrix for f.

2. If A is an $n \times n$ matrix over a field F satisfying $A^n = 0$, $A^{n-1} \neq 0$, find the rational and Jordan canonical matrices similar to A.

3. If K is a field containing a field F and $A \in F^{n \times n}$, show that the invariant factors and minimum polynomial of A as $n \times n$ matrix over F are the same as the invariant factors and minimum polynomial of A as $n \times n$ matrix over K.

Find the elementary divisors of $\begin{pmatrix} 0 & 1 \\ 2 & 0 \end{pmatrix}$

     (i) as element of $\mathbb{Q}^{2 \times 2}$;
     (ii) as element of $\mathbb{R}^{2 \times 2}$.

4. If $A \in F^{n \times n}$, prove that A^t is similar to A. (Use Proposition 10 and exercise 8 of § 4.6, p. 83. For another method of proof see the next exercise.)

5. If B is the matrix of Lemma 1 and if

$$P = \begin{pmatrix} d_{s-1} & d_{s-2} & \cdots & & d_1 & -1 \\ d_{s-2} & d_{s-3} & \cdots & d_1 & -1 & 0 \\ \cdot & \cdot & \cdot & \cdot & \cdot & \cdot \\ d_1 & -1 & 0 & \cdots\cdots & & 0 \\ -1 & 0 & 0 & \cdots\cdots & & 0 \end{pmatrix}$$

show that $PB = B^t P$. Deduce from Theorem 3 that any square matrix is similar to its transposed matrix.

7.6 Classification of Collineations

Let f be a linear mapping of V into V and suppose that the elementary divisors of f are $(X - x_i)^{n_{ij}}$, $i = 1, \ldots, t$; $j = 1, \ldots, k_i$, where

$$n_{i1} \geq n_{i2} \geq \cdots \geq n_{ik_i}.$$

To each eigenvalue x_i there corresponds the k_i-tuple $(n_{i1}, \ldots, n_{ik_i})$ and we define the set of all these,

$$\{(n_{11}, \ldots, n_{1k_1}), \ldots, (n_{t1}, \ldots, n_{tk_t})\},$$

to be the *Segre symbol* of f. We stress the fact that the Segre symbol is defined only for linear mappings whose minimum polynomials factorize in $F[X]$ into linear factors. Of course, if $F = \mathbf{C}$, then this is true for all linear mappings of V into V.

The importance of the Segre symbol lies in the classification of collineations. If the elementary divisors of f are $(X - x_i)^{n_{ij}}$, then the elementary divisors of cf are $(X - cx_i)^{n_{ij}}$, for any $c \neq 0$ in F. Hence the Segre symbols of f and cf are the same. We may therefore define unambiguously the Segre symbol of $\mathscr{P}(f)$ to be that of f.

Let f be an automorphism of V. If we call the configuration of f-invariant subspaces of $\mathscr{P}(V)$ the *invariant configuration* of the collineation $\mathscr{P}(f)$, then we may say that the geometrical character of $\mathscr{P}(f)$ is summarized by being given its invariant configuration. The precise meaning of this statement is the content of the next result.

THEOREM 5. *Let V be a vector space over F and suppose that f, g are automorphisms of V whose minimum polynomials are products of linear factors in $F[X]$. Then the collineations $\mathscr{P}(f)$, $\mathscr{P}(g)$ have the same Segre symbol if, and only if, there is a collineation of $\mathscr{P}(V)$ mapping the invariant configuration of $\mathscr{P}(f)$ onto the invariant configuration of $\mathscr{P}(g)$.*

PROOF. (i) Suppose the Segre symbols are the same and let the elementary divisors of f, g be $(X-x_i)^{n_{ij}}$, $(X-y_i)^{n_{ij}}$, respectively. Suppose

$$V_f = \bigoplus_{i=1}^{t} K_i, \qquad V_g = \bigoplus_{i=1}^{t} L_i,$$

where, for each i, K_i is the submodule of V_f consisting of all elements annihilated by some power of $X-x_i$; and L_i similarly for $X-y_i$ in V_g. If f_i, g_i are the restrictions of f, g to K_i, L_i, respectively, then the Segre symbol of both f_i and g_i is $\{(n_{i1}, \ldots, n_{ik_i})\}$.

Since f_i and $(x_i/y_i)g_i$ have the same elementary divisors, there is a module isomorphism e_i of K_i onto L_i. Clearly, e_i must map the invariant configuration of $\mathscr{P}(f_i)$ onto the invariant configuration of $\mathscr{P}((x_i/y_i)g_i) = \mathscr{P}(g_i)$. If e is the result of combining $e_1, \ldots, e_t$, then e is an isomorphism of V_f onto V_g.

Now any submodule M of V_f has a direct decomposition

$$M = (M \cap K_1) \oplus \cdots \oplus (M \cap K_t)$$

where $M \cap K_i$ is the set of all elements of M annihilated by some power of $(X-x_i)$. There is a similar decomposition for any submodule of V_g. It follows that e maps the invariant configuration of $\mathscr{P}(f)$ onto that of $\mathscr{P}(g)$ and thus $\mathscr{P}(e)$ is the required collineation.

(ii) Since f and $e^{-1}fe$ have the same Segre symbol (Proposition 5) we may assume without loss of generality that $\mathscr{P}(f)$ and $\mathscr{P}(g)$ have the same invariant configuration. Let

$$V_f = \bigoplus [a_{ij}]$$

be the decomposition (according to Theorem 2) into indecomposable submodules, where

$$a(a_{ij}) = [(X-x_i)^{n_{ij}}].$$

By our hypothesis on $\mathscr{P}(f)$ and $\mathscr{P}(g)$, each $[a_{ij}]$ must be an indecomposable submodule of V_g and hence is annihilated by some power of an irreducible factor of the minimum polynomial of g. The uniqueness part of Theorem 2 applied to V_g now shows that the numbers $\dim_F[a_{ij}]$ which appear in the Segre symbol of f are precisely those which appear in the Segre symbol of g. But two invariant points of a collineation are joined by a line of invariant points if, and only if, they correspond to the same eigenvalue. Since the collineations $\mathscr{P}(f)$, $\mathscr{P}(g)$ have the same invariant configuration this shows finally that the Segre symbols of $\mathscr{P}(f)$ and $\mathscr{P}(g)$ are the same.

The invariant configuration of a collineation has the important property of being self-dual. This is the content of our final result.

THEOREM 6.   *Given a collineation $\mathscr{P}(f)$ of $\mathscr{P}(V)$, there exists a correlation of $\mathscr{P}(V)$ mapping the invariant configuration of $\mathscr{P}(f)$ onto itself.*

PROOF.   By Proposition 10, there is a module isomorphism g of V_f onto $(V^*)_{f^t}$. We show that the correlation $\mathscr{P}(g)\circ$ has the required property.

For any v in V we have

$$vfg = (Xv)g = X(vg) = vgf^t.$$

Hence, for any v, w in V,

$$(vf)(wg) = v(wgf^t), \quad \text{by the definition of } f^t,$$
$$= v(wfg), \quad \text{by the last equation.}$$

Suppose M is f-invariant.   Then for all v in $(Mg)^\circ$ and all w in M,

$$v(wfg) = 0,$$

whence

$$(vf)(wg) = 0.$$

Thus $(Mg)^\circ f \subset (Mg)^\circ$ and so $(Mg)^\circ$ is f-invariant.   On the other hand, if $(Mg)^\circ$ is f-invariant, then

$$(vf)(wg) = 0$$

for all v in $(Mg)^\circ$ and all w in M and therefore

$$v(wfg) = 0.$$

This shows that $Mfg \subset (Mg)^{\circ\circ} = Mg$ and so $Mf \subset M$, i.e., M is f-invariant.

Theorem 6 can often be used to discuss a given collineation without reference to the theory of elementary divisors; and in fact can be used when the Segre symbol is not defined so that Theorem 5 is not available.

Suppose, for example, that a collineation π of a 3-dimensional projective geometry P over a field F has precisely two invariant points A, B.   Then we know at once that π has exactly two invariant planes P, Q, say.   Clearly, the line L joining A, B is an invariant line, as also is $P \cap Q$.   There are *at most* three possibilities:

   (i) L does not lie in P or Q;
   (ii) L lies in P, say, but not in Q;
   (iii) $L = P \cap Q$.

In case (i) L meets each of the planes P, Q in an invariant point.   We may suppose that $L \cap P = A$, $L \cap Q = B$.   By considering the restriction of π to P and using Theorem 6 again we see that there is just one invariant line of π in P, viz. $P \cap Q$; and similarly for Q.   Any invariant line M which is not in P or Q must intersect P, Q in invariant points and so $M = L$. The invariant configuration therefore consists of two points A, B, two planes P, Q and two lines AB, $P \cap Q$.

Case (ii). Since L meets Q in an invariant point we may suppose that B lies on $P \cap Q$. Using Theorem 6 we see that there is just one invariant line in Q, viz. $P \cap Q$. Any invariant line M which is not in P or Q must intersect Q at the invariant point B; but then the join of M and A would give a third invariant plane. Thus L and $P \cap Q$ are the only invariant lines.

Case (iii). Since P contains A and B, there are just two invariant lines in P, viz. L and one other, say M, which meets L in an invariant point. We may suppose that $M \cap L = A$. Similarly, there is an invariant line N in Q that cannot go through A as we should then have a third invariant plane joining M and N. Thus N goes through B. We at once check that L, M, N are the only invariant lines in this case.

We observe that if P is defined over the field C of complex numbers (or more generally over any algebraically closed field) then by Proposition 3 of § 6.3, p. 135, any collineation has invariant points. In particular the restriction of π to $P \cap Q$ in the above example must have invariant points so that case (i) cannot occur. All three cases do occur for example over the field R of real numbers: the matrices

$$\begin{pmatrix} 1 & 0 & & \\ 0 & -1 & & \\ & & 0 & 1 \\ & & -1 & 0 \end{pmatrix}, \quad \begin{pmatrix} 2 & 1 & & \\ & 2 & 1 & \\ & & 2 & \\ & & & 3 \end{pmatrix}, \quad \begin{pmatrix} 2 & 1 & & \\ & 2 & & \\ & & 3 & 1 \\ & & & 3 \end{pmatrix},$$

define collineations of $\mathscr{P}(R^4)$ of types (i), (ii) and (iii), respectively.

The reader should observe that the simplest matrices describing a given collineation are obtained by choosing a simplex of reference so that as many as possible of its vertices, sides, faces, etc. belong to the invariant configuration.

EXERCISES

1. Find the Segre symbols and invariant configurations for the non-identity collineations of an n-dimensional projective geometry over C when $n = 1$, 2, 3. (For $n = 1$ there are two cases; for $n = 2$ there are five; for $n = 3$ there are thirteen.)

2. Use Theorem 6 to find the invariant configurations for the non-identity collineations of a real projective plane.

3. Find the invariant configuration of a collineation of a 3-dimensional projective geometry given that it has just three invariant points. (Assume that the ground field has more than two elements.)

4. Let P be a real projective geometry of dimension 3. Show that a collineation of P without any invariant points has one, two or infinitely many invariant lines.

5. Let F be a finite field with p^{n+1} elements and let a be a generator of the cyclic group F^*. (See exercise 11 of § 7.3.) Prove that

$$f: x \rightarrow xa$$

is an automorphism of F, viewed as a vector space over $\mathbf{F}_p$, and that f has order $p^{n+1} - 1$. Deduce that $\mathscr{P}(f)$ is a collineation of $\mathscr{P}(F)$ of order $1 + p + \cdots + p^n$.

(Observe that $1 + p + \cdots + p^n$ is the number of points in $\mathscr{P}(F)$ (cf. exercise 5 of § 2.5, p. 32). Now, by Theorem 6 of Chapter II, p. 32, any projective geometry P of dimension n over $\mathbf{F}_p$ is isomorphic to $\mathscr{P}(F)$. Hence our exercise implies that there exists a collineation that permutes all the points of P cyclically. In fact this result is true for all projective geometries over $\mathbf{F}_p$ because there exist fields with p^{n+1} elements for all $n \geq 0$.)

§ 2.1

4. Let c be an element of S and consider the set $M = -c + S$. Then M has the property of exercise 3 and also contains the zero vector. If $a, b \in M$ and $x, y \in F$ then (by hypothesis, with $r = 3$)

$$xa + yb = xa + yb + (1 - x - y)0 \in M$$

and so M is a subspace.

5. Consider $M = -c + S$ as in 4. If $a \in M$ and $x \in F$ then

$$xa = xa + (1 - x)0 \in M.$$

By hypothesis we can find non-zero scalars y, z for which $y + z = 1$. Then

$$a + b = y\left(\frac{1}{y}a\right) + z\left(\frac{1}{z}b\right) \in M$$

for any a, b in M.

The non-zero vectors in $\mathbf{F}_2{}^2$ do not form a coset.

§ 2.6

2. Let P, Q be elements of A with hyperplanes at infinity M, N, respectively, i.e., $M = P \cap H, N = Q \cap H$.

THEOREM 1. *Let $P, Q \in \mathsf{A}$.*
(1) *If $P \subset Q$, then* pdim $P \le$ pdim Q *and* pdim $P =$ pdim Q *implies* $P = Q$.
(2) *If $P \cap Q \in \mathsf{A}$, then*

$$\text{pdim}(P + Q) + \text{pdim }(P \cap Q) = \text{pdim } P + \text{pdim } Q.$$

(3) (i) *Assuming $P \cap Q \in \mathsf{A}$; $M \subset N$ if, and only if, $P \subset Q$.*
(ii) *Assuming $P \cap Q \notin \mathsf{A}$; $M \subset N$ if, and only if,*

$$\text{pdim}(P + Q) = \text{pdim } Q + 1.$$

PROOF: (1) and (2) are immediate by Theorem 2 of Chapter I.
(3) (i). If $P \cap Q \in \mathsf{A}$ there exists a vector c in $P \cap Q$ but not in H and so

$$P = [c] + M, \qquad Q = [c] + N.$$

(ii). $P \cap Q \subset H$. If $M \subset N$ then $M \subset P \cap Q \subsetneq P$. Counting dimensions gives $M = P \cap Q$ and so

$$\text{dim}(P + Q) = \text{dim } P + \text{dim } Q - \text{dim } M = \text{dim } Q + 1.$$

175

Conversely, if $\dim(P+Q)=\dim Q+1$ then $\dim(P\cap Q)=\dim P-1$. But $P\cap(\bar{P}\cap Q)\subset P\cap H$ gives $P\cap Q\subset M$, so that $P\cap Q=M$ and $M\subset N$.

§ 3.2

4. Q is any point not on L or PP'. Let $B=L\cap PQ$ and $Q'=AQ\cap BP'$. Case (i): A not on L. If $P_0=PP'\cap L$, $Q_0=QQ'\cap L$, then there is a unique collineation taking (P, Q, P_0, Q_0) to (P', Q', P_0, Q_0) (Theorem 2). Case (ii): A on L. If C is any point on L different from A and B, there is a unique collineation taking (P, Q, A, C) to (P', Q', A, C). These collineations are central as required.

5. $B\pi$ must lie on PB and on LC' and so $B\pi=B'$. Similarly, $A\pi=A'$. Now $(AB)\pi$ is a line through $AB\cap LM$, i.e., N lies on LM.

6. Draw any line M through $L\cap L'$ different from L and L'. Use exercise 4 to find a central collineation with axis M, center A and mapping L onto L'.

8. Let P, Q, R be distinct points on L and P', Q', R' their images on L'. We may assume that P, P' are distinct from $L\cap L'$. Let then $Q''=PQ'\cap P'Q$ and $R''=PR'\cap P'R$. Consider the perspectivity of L onto $Q''R''$ with center P' and the perspectivity of $Q''R''$ onto L' with center P.

10. *Last part.* If $zx\neq xz$ and if $(x_1, x_2, x_3)f=(x_1z, x_2z, x_3z)$, then $\mathscr{P}(f)$ is not the identity collineation on $\mathscr{P}(D^3)$: e.g., $[(1, x, 0)]\to[(z, xz, 0)]$.

§ 3.5

1. If ζ is an automorphism of $\mathbf{Q}$, $1\zeta=1$ and so $n\zeta=n$ for all $n\in\mathbf{Z}$ (by additivity). Further,

$$n\left(\frac{m}{n}\zeta\right) = m\zeta \quad \text{gives} \quad \frac{m}{n}\zeta=\frac{m}{n}.$$

If ζ is an automorphism of $\mathbf{C}$ and $\mathbf{R}\zeta=\mathbf{R}$, then

$$(a+ib)\zeta = (a\zeta)+(i\zeta)(b\zeta) \quad \text{and} \quad (i\zeta)^2 = -1.$$

So $i\zeta=\pm i$.

3. If $z(x_1, \ldots, x_n)=(\bar{x}_1, \ldots, \bar{x}_n)$ then $|z|=1$ and if $z=e^{2i\theta}$,

$$e^{i\theta}(x_1, \ldots, x_n) = e^{-i\theta}(\bar{x}_1, \ldots, \bar{x}_n)$$

is a real n-tuple.

§ 3.6

1. If $1\varphi=0$ then all $x\varphi=0$. Assuming $1\varphi\neq0$; if $x\neq0$, then $(x\varphi)(x^{-1}\varphi)=1\varphi$ implies $x\varphi\neq0$.

§ 4.1

6. For given i, j

$$f_{ij}: \sum x_k u_k \to x_i v_j.$$

If $u_i f = \sum_j a_{ij} v_j$ then f is uniquely expressed as $\sum_{i,j} a_{ij} f_{ij}$.

11. If $c \notin H$, every vector is uniquely expressible as $xc + h$. Now $\bar{f}$ is the identity if, and only if, $vf - v \in H$ for all v, i.e., if, and only if, $cf - c \in H$. But $[cf - c]$ is the center of $\mathscr{P}(f)$.

§ 4.2

2. $\operatorname{Ker} f = \operatorname{Ker} g$. If $f \neq 0$ and M is a direct complement in V to $\operatorname{Ker} f$, apply Proposition 1 of Chapter III (p. 45) to the restrictions of f, g to M.

3. $\mathscr{P}(g^2) = \mathscr{P}(g)$ implies $g = zg^2$, by exercise 2. Put $f = zg$.
   If P is any subspace then $Pf = (K + P) \cap M$ where K, M, respectively, are the kernel and image of f. It is natural to call $\mathscr{P}(f)$ the "projection from K onto M".

§ 4.3

9. If $i \neq j$, $0 = (E_{kj} E_{ik})f = (E_{ik} E_{kj})f = E_{ij}f$, and $E_{ii}f = (E_{ij} E_{ji})f = (E_{ji} E_{ij})f = E_{jj}f$.

§ 4.4

2. The given condition is $x(AA^t - I)x^t = 0$ for all x. Now $xSx^t = 0$ for all x is equivalent to $s_{ii} = 0$, all i, and $s_{ij} = -s_{ji}$ all $i \neq j$. If F is not of characteristic 2 this is equivalent to the skew-symmetry of S. Hence then $x(AA^t - I)x^t = 0$ for all x is equivalent to $2(AA^t - I) = 0$, i.e., $AA^t = I$. But if F is of characteristic 2, $x(AA^t - I)x^t = 0$ for all x is equivalent to $\sum_k a_{ik}^2 = 1$ for all i.

5. If $C = AB$ with $r(A) = r(B) = r$, then $r(C) = r$ by Proposition 6 (3) and exercise 4.
   Conversely, if $r(C) = r$, then C is equivalent to

$$\begin{pmatrix} I_r & 0 \\ 0 & 0 \end{pmatrix}_{m \times r} = \begin{pmatrix} I_r \\ 0 \end{pmatrix} \begin{pmatrix} I_r & 0 \end{pmatrix}_{r \times p}$$

§ 4.6

5. *Last part.* Let φ be the linear form $p(X) \to p(1)$. If $\varphi = a_0 f_0 + \cdots + a_k f_k$, then $X^{k+1} \varphi = 0$, a contradiction.

6. Clearly ι is linear. If $p(X)\iota = 0$ then $p(X)f_i = 0$ for all $i \geq 0$, where $f_0, f_1, \ldots$ are the linear forms defined in exercise 5. Hence $p(X) = 0$.

There exists an element φ of V^{**} mapping every element of the basis of V^* to 1. If $\varphi = p(X)\iota$ where $p(X)$ has degree n, then we obtain $f_{n+1}\varphi = p(X)f_{n+1} = 0$, a contradiction.

(The reader who is familiar with the use of Zorn's Lemma will easily verify that the linearly independent subsets of V^* containing $\{f_0, f_1, \ldots\}$ are inductively ordered by inclusion. A maximal subset of this collection is a basis of V^*.)

7. To show that $(f^t)^t$ is identified with f amounts to proving $(u\iota)(f^t)^t = (uf)\iota$, all u in U. Now $(u\iota)(f^t)^t = f^t(u\iota)$ (definition of transpose). If $h \in V^*$, $(hf^t)(u\iota) = u(hf^t)$ (definition of ι) and $hf^t = fh$. So $(hf^t)(u\iota) = (uf)h = h(uf\iota)$, i.e., $f^t(u\iota) = (uf)\iota$.

§ 4.8

1. (1) The equation of U' with respect to (A_0, A_1, U) is $X_0 + X_1 = 0$. So U' is the point $(1, -1)$ in $\mathscr{P}(V)$.

(2) If $(a_0, \ldots, a_n)$ determines $(A_0, \ldots, A_n, U)$, then $(a_0, \ldots, a_{n-1})$ determines $(A_0, \ldots, A_{n-1}, U_n)$. If $p_0X_0 + \cdots + p_nX_n = 0$ is the equation of a hyperplane P of $\mathscr{P}(V)$, then $p_0X_0 + \cdots + p_{n-1}X_{n-1} = 0$ is the equation of $P \cap A_n'$ with respect to $(A_0, \ldots, A_{n-1}, U_n)$.

(3) When $n = 1$ use (1). Inductive step: $U' = +(U' \cap A_i' : i = 0, \ldots, n)$.

§ 5.1

1. The statement $\sigma(a, b) = 0$ for all b is equivalent to $a\underset{\sim}{g} = 0$. Thus (i) is equivalent to g having kernel 0. Similarly for (ii).

The statement $\sigma(a, b) = 0$ for all a is equivalent to $b \in (Vg)^\circ$. In the finite dimensional case $(Vg)^\circ = 0$ if, and only if, $Vg = V^*$.

§ 5.2

4. For any subspace N, $V^\perp \subset N^\perp$. Hence $V^\perp \subset M^\perp \cap (M^\perp)^\perp = 0$.

§ 5.3

9. (4). Here $\alpha\beta = \beta\alpha = 0$ and so Image $\alpha \subset$ Ker β, Image $\beta \subset$ Ker α.

Let $(a_1, \ldots, a_n)$ be an ordered basis of V. If $\sigma\alpha = 0$, let q be the quadratic form $q(\sum x_i a_i) = \sum_{i < j} \sigma(a_i, a_j)x_i x_j$. Then $\sigma = q\beta$.

If $q\beta = 0$, let σ be the symmetric bilinear form

$$\sigma(\sum x_i a_i, \sum y_i a_i) = \sum_i q(a_i)x_i y_i .$$

Then $q = \sigma\alpha$ (by exercise 8).

§ 5.4

3. $\sigma(a + xc, a + xc) = \sigma(a, a) + 2x\sigma(a, c) \geq 0$ for all x.

7. Let σ be the bilinear form on $\mathbf{F}_2^3$ having the given matrix with respect to the standard basis. Now σ is non-degenerate and so, if we could

triangularize σ, there would be three vectors with $\sigma(a, a) \neq 0$. But if $a = (x, y, z)$, $\sigma(a, a) = x(x+y)$ and so the only such vectors are $(1, 0, 0)$, $(1, 0, 1)$.

8. There is a smallest positive integer p such that $p1_F = 0$ and p must be prime. Then $\{m1_F; m = 0, 1, \ldots, p-1\}$ is a subfield isomorphic to $\mathbf{F}_p$. E is a finite dimensional vector space over F_0.

9. Since F is finite, $S^* = F^*$ if, and only if, the kernel is $\{1\}$; and this holds if, and only if, $1 = -1$, i.e., $2 = 0$.

If $1 \neq -1$, each element in S^* is the image of exactly two elements in F^*. Thus the number of elements in F^* is twice the number in S^*. Now $\{zs; s \in S^*\}$ has empty intersection with S^* and contains the same number of elements as S^*.

10. If $x, y \in S$ but $x + y \notin S$ then x, y must be non-zero and $x + y = zs$ for some s in S^*. This contradicts the hypothesis on z. Hence $x + y \in S$ and S is closed with respect to addition. In particular, $(p-1)x \in S$, i.e., $-x \in S$. Obviously, S is closed with respect to multiplication and S^* is closed with respect to inversion.

The number of elements in S is $1 + \frac{1}{2}(q-1) = \frac{1}{2}(q+1)$ and this cannot be a power of p because q is a power of p.

11. Let $(\sigma; (a_1, \ldots, a_n)) = \operatorname{diag}(x_1, \ldots, x_r, 0, \ldots, 0)$. Each x_i is either in S^*, say $x_i = y_i^2$, or has the form $x_i = zy_i^2$ (exercise 9). Then $\sigma(y_i^{-1}a_i, y_i^{-1}a_i) = 1$ or z. So we obtain an ordered basis $(b_1, \ldots, b_n)$ with respect to which σ has matrix $\operatorname{diag}(1, 1, \ldots, 1, z, z, \ldots, z, 0, \ldots, 0)$. The z's can now be knocked off in pairs. Let $z^{-1} = z_1^2 + z_2^2$ (exercise 10). If

$$\sigma(b_i, b_i) = \sigma(b_j, b_j) = z$$

put

$$b_i' = z_1 b_i + z_2 b_j,$$
$$b_j' = -z_2 b_i + z_1 b_j$$

and check that $\sigma(b_i', b_i') = \sigma(b_j', b_j') = 1$, $\sigma(b_i', b_j') = 0$.

§ 5.5

3. $P_i \subset P_i^\perp$ and so the vertices of the first tetrahedron lie on the faces of the second. On the other hand,

$$P_2^\perp \cap P_3^\perp \cap P_4^\perp = (P_2 + P_3 + P_4)^\perp \subset (P_2 + P_3 + P_4)^{\perp\perp} = P_2 + P_3 + P_4,$$

where the left-hand side is a vertex of the second tetrahedron and the right-hand side a face of the first.

4. Let $\perp$ be a possibly degenerate null-polarity. If $P \subset Q^\perp$ then we have $P \subset P^\perp \cap Q^\perp = L^\perp$. But $P \subset Q^\perp$ also implies $Q \subset P^\perp$ and so $L \subset L^\perp$. (In the non-degenerate case $\dim L = \dim L^\perp$ and so $L = L^\perp$.) Assume conversely that $L \subset L^\perp$. Then for any points P, Q on L we have $P \subset L \subset L^\perp \subset Q^\perp$.

The lines through P which satisfy $L \subset L^\perp$ are simply the lines through P in $P^\perp$. When $\perp = \perp(\sigma)$ and σ is degenerate but not zero, $V^\perp$ is a line; then $P^\perp$ is the plane $P + V^\perp$ for any point P not on $V^\perp$. The linear complex in this case consists of $V^\perp$ and all the lines meeting $V^\perp$.

5. B consists of the subspaces M of V for which both M and $M^\perp$ lie in A. Hence $\perp'$ is a mapping of B into B. Since $\perp'\perp'$ is the identity mapping it follows that $\perp'$ is one–one and onto B.

Knowing only $\perp'$ we try to reconstruct $\perp$. It will be sufficient to have the image of each point in $\mathscr{P}(V)$: for then the formula $(M+N)^\perp = M^\perp \cap N^\perp$ will yield the image of each element of $\mathscr{P}(V)$. If there is a point of $\mathscr{P}(V)$ in A but not in B, then this point *must* be $H^\perp$ and so its image must be H. This has extended $\perp'$ to all points in A.

Now take a point P in H. Let L be a line so that $L \cap H = P$. For any point $Q \neq P$ on L, $Q^\perp$ is known and contains $L^\perp$. As Q varies, we obtain all hyperplanes $Q^\perp$ containing $L^\perp$ except for one: this exception must be $P^\perp$.

6. B consists of the subspaces M for which M and $M^\perp$ are both in A.

If $V^\perp \in$ A then $M^\perp \in$ A for all M (because $M^\perp \supset V^\perp$). Thus $B=$A. Conversely, if $B=$ A then $V \in B$ and so $V^\perp \in$ A.

Assume $V^\perp = H$. Then $M^\perp \supset H$ for all M. Either $M^\perp = H$ or $M^\perp = V$ which would imply $M \subset V^\perp = H$. In any case $M \notin B$. Conversely, assume B empty. For any point P in A, $P^\perp \subset H$ and thus $P^\perp = H$ because dim $P^\perp \geq$ dim $V-1$ (Proposition 4, p. 93). Now V may be expressed as the join $P_0 + \cdots + P_n$ of points P_i in A and so finally $V^\perp = P_0^\perp \cap \ldots \cap P_n^\perp = H$.

Clearly $\perp'$ is a mapping of B into A. We wish to reconstruct $\perp$ from $\perp'$. We need only find $P^\perp$ for all *points* P (cf. solution to exercise 5, above).

If $P \in$ A but $P \notin B$, then $P^\perp \subset H$. But $P^\perp$ is a hyperplane or is V. Hence $P^\perp = H$.

Consider a line L in $\mathscr{P}(V)$ and a variable point P on L. There are three cases: (i) all $P^\perp$ are distinct hyperplanes; (ii) one $P^\perp$ is V and all the others are equal to a fixed hyperplane; (iii) all $P^\perp = V$. (If $\perp = \perp(\sigma)$, these cases occur according as $L\sigma$ has dimension 2, 1 or 0, respectively.) Since we know $P^\perp$ for every point P in A we can find $P^\perp$ for any point P in H by choosing a line L as in exercise 5 which has P as its point at infinity.

§ 5.6

6. $(1, 1, 1)^\perp$ meets $X_2 = 0$ at $(1, -1, 0)$. If $m+n \neq 0$, find the unique point, other than $(1, 1, 1)$, at which the line joining $(1, 1, 1)$ and $(m, n, 0)$ meets the conic.

7. Let $A \in Q(\sigma)$. Since there are at least three lines through A, there is a line L distinct from $A^\perp$ and AP. Hence L meets Q in $B \neq A$ (exercise 5) and $PB \neq PA$.

If $AP = A^\perp$ then $P^\perp$ goes through A; in the other case $P^\perp$ goes through the harmonic conjugate of P with respect to the point pair $AP \cap Q$. Similarly for BP.

If $P \in Q$, $P^\perp$ is the unique line cutting Q only at P.

10. By the last part of exercise 5, $Q(\sigma)$ contains at least three points, say A_0, A_2, E. Now use exercises 8, 9.

12. If the signature is ± 2, then the equation of the quadric in a suitable co-ordinate system is

$$X_0{}^2 + X_1{}^2 + X_2{}^2 - X_3{}^2 = 0.$$

If there were a generator, the plane $X_3 = 0$ would have to meet it in a point (or points) satisfying $X_0{}^2 + X_1{}^2 + X_2{}^2 = 0$.

13. Let the skew lines be A, B, C. Choose distinct points A_1, A_3 on A and let the unique transversal from A_1 to B, C meet B at A_0 and the unique transversal from A_3 to B, C meet B at A_2. Choose a point U on C but not on either of these transversals. Then A, B, C have the required equations with respect to the frame of reference (A_0, A_1, A_2, A_3, U).

The transversal from $(0, x, 0, y)$ on A meets B at $(x, 0, y, 0)$ and C at (x, x, y, y).

14. Since L is a generator $L^\perp = L$. If M is a plane containing L then the point $M^\perp$ lies on $L^\perp = L$. Now $M \cap M^\perp = M^\perp$ and so σ_M has rank 2, i.e., the conic $Q \cap M$ is a pair of distinct lines through $M^\perp$ (one of which is L). Any three distinct planes through L yield three skew lines A, B, C which are generators of Q. If we choose a frame of reference as in exercise 13 and with $A_2 A_3 = L$ then Q and L have the required equations.

16. (i) If $n = 1$, the hypothesis forces σ to be non-degenerate (exercise 2) and then we use exercise 3.

(ii) Q cannot be contained in a hyperplane. Suppose $Q \subset H$, H a hyperplane. Take any point A on Q and any line L containing A but $L \not\subset H$. Then L meets Q only in A and hence σ_L is degenerate. Thus $\sigma(a, b) = 0$ for all $b \notin H$. Each vector in H can be written as a sum of two vectors not in H. Therefore $\sigma(a, v) = 0$ for all v in V and thus $A \subset V^\perp$. So $Q \subset V^\perp$, a contradiction.

(iii) As $P \notin Q$, $P^\perp \neq V$ and so $P^\perp$ is a hyperplane. By (ii), $Q \not\subset P^\perp$ and we can find A in Q such that $A \not\subset P^\perp$, i.e., $P \not\subset A^\perp$. Take H to be any hyperplane containing A and P, and therefore distinct from $A^\perp$. Then the quadric $Q \cap H$ satisfies the same condition in H as Q does in V $(A^{\perp(\sigma_H)} \neq H)$.

By induction on n, there exist $n-1$ lines through P, each meeting $Q \cap H$ and whose join is H. Now $Q \not\subset H$ by (ii) and so there exists a point B in Q not in H. Then PB meets Q and $PB + H = V$.

(Remark. If the ground field is $\mathbf{R}$, the quadric has equation

$$-X_0{}^2 + X_1{}^2 + X_2{}^2 = 0$$

and P has coordinates $(1, 0, 0)$, then the polar of P does not meet the quadric.)

(iv) Let $A_1, \ldots, A_n$ be the points of Q such that $V = PA_1 + \cdots + PA_n$ (by (iii)). If PA_i meets Q in a second point B_i, define P_i to be the harmonic conjugate of P with respect to A_i, B_i; if PA_i meets Q only at A_i, define P_i to be A_i. Then $P_1 + \cdots + P_n$ is a hyperplane contained in $P^\perp$ and hence equals $P^\perp$.

(v) Let $P \in Q$. If σ_L is degenerate for every line L through P, $P \subset V^\perp$. If there is a line L containing P such that σ_L is non-degenerate, then L certainly contains two points not on Q. Hence $L^\perp$ is known by (iv). Now P must be $P + L^\perp$.

§ 5.7

3. $Q(\sigma)$ lies in the hyperplane H. Let $A \in Q(\sigma)$ and B be any point not in H. AB meets Q only at A and so σ_{AB} is degenerate. Thus A is orthogonal to B. But V is spanned by the points B not in H and so $A \subset V^{\perp}$.

4. Rank 0 is always possible. Let $V = M \oplus V^{\perp}$ where $\dim M = \operatorname{rank} \sigma = r \geq 1$ and $Q(\sigma_M)$ is empty. Choose coordinates so that Q has equation $a_0 X_0{}^2 + \cdots + a_{r-1} X_{r-1}^2 = 0 \ (a_0 \cdots a_{r-1} \neq 0)$.

(i) As long as $a_0, \ldots, a_{r-1}$ are all positive or all negative $Q(\sigma_M)$ is empty. Thus any $r \geq 1$ is possible.

(ii) If $r > 1$, $Q(\sigma_M)$ is not empty. Hence $r = 1$.

(iii) If $r > 2$, $Q(\sigma_M)$ is not empty. If $r = 2$ and $a_0 = a_1$, then $Q(\sigma_M)$ is empty. Hence r is 1 or 2.

5. Let $Q \subset H \cup K$ (H, K two distinct hyperplanes). Since Q is non-empty and non-degenerate, $Q \not\subset H$ and $Q \not\subset K$ by exercise 3. We may assume pdim $V > 1$. Thus $H \cap K$ contains points.

(i) $Q \cap H \cap K$ *is empty*. For if P is a point of Q in $H \cap K$ and R is any point not in H or K, PR meets Q only at P and so $R \subset P^{\perp}$ (where $\perp$ is any polarity determining Q). Thus $P^{\perp} = V$ (cf. solution to exercise 3), i.e., $V^{\perp} \neq 0$, a contradiction.

(ii) Let P be a point of Q in H. Then $P^{\perp} \neq K$ by (i) and so $P^{\perp} \cap K$, $H \cap K$ are hyperplanes in K. Any line through P, not in H and not in $P^{\perp}$, must meet Q in one more point P' and $P' \subset K$. Hence $Q \cap K$ *contains all points in K that are not in H and not in $P^{\perp}$*.

(iii) If pdim $V > 2$, $(H \cap K) \cap (P^{\perp} \cap K)$ contains points. Let R be one such. Choose any point R' in K but not in H or $P^{\perp}$. Then R is the only point of the line RR' in H and also the only point of the line in $P^{\perp}$. Then Q contains all the points of RR' except R (by (ii)). Since the ground field contains at least three elements, Q contains the whole line RR', contrary to (i). Hence pdim $V \leq 2$.

(iv) Let pdim $V = 2$. Here $H \cap K$, $P^{\perp} \cap K$ are points on the line K. If K contains more than four points then by (ii) Q contains all the points of K, a contradiction to (i). Hence K contains exactly four points, i.e., the ground field is (isomorphic to) $\mathbf{F}_3$. Furthermore, K (and similarly H) contains just two points of Q, so that Q consists of four points. (Cf. exercise 5 of § 5.6.)

6. Let $Q \subset H \cup K$, $H \neq K$.

(i) We may assume $Q \not\subset H$ and $Q \not\subset K$, otherwise use exercise 3.

(ii) If A is a point in $V^{\perp}$, $B \in Q$ and $A \neq B$, then $AB \subset Q$.

(iii) $V^{\perp} \subset H \cap K$. For if A is a point in $V^{\perp}$ not in $H \cap K$, say $A \not\subset K$, then there is a point, say B, of Q not in H (by (i)) and $AB \subset Q$ by (ii). But AB contains points not in $H \cup K$, a contradiction.

(iv) Let $V = M \oplus V^{\perp}$, so that $Q \cap M$ is a non-degenerate quadric in $\mathscr{P}(M)$. By (iii) and (i), $M \not\subset H$ and $M \not\subset K$. So $H \cap M$, $K \cap M$ are hyperplanes in M. Since Q is not contained in $V^{\perp}$ (by (iii) and (i)) we see by exercise 17 of § 5.6 that $Q \cap M$ is not empty and so is one of the two types of quadric given in exercise 5 above. We get Q from $Q \cap M$ by means of exercise 17 of § 5.6: if $Q \cap M$ is a pair of points then Q

consists of all the points on the hyperplanes H and K; in the second case Q consists of the points on four subspaces, two being hyperplanes in H and two hyperplanes in K.

7. Let A be determined by H in $\mathscr{P}(V)$. Let Q' be an affine quadric contained in a hyperplane K of A and let Q be a projective quadric completing Q'. Now $Q \subset H \cup K$. If $Q \subset H$ then Q' is empty and if $Q \subset K$ then Q' consists of all the affine points in some subspace N of K (using exercise 3). If Q consists of all the points of $H \cup K$ then Q' consists of all the affine points of K. (Note that $Q = H \cup K$ and $Q = K$ give the same Q'.) By exercise 6 there is only one further case: Q' is the set of all affine points in a pair of hyperplanes of K and the ground field is isomorphic to $\mathbf{F}_3$.

8. Over $\mathbf{F}_3$ there exists a unique non-degenerate conic through any four points, no three of which are collinear. (Thus $X_0 X_1 + X_1 X_2 + X_2 X_0 = 0$ is the unique non-degenerate conic through $(1, 0, 0)$, $(0, 1, 0)$, $(0, 0, 1)$, $(1, 1, 1)$.)

If a completion were degenerate it would be a line pair and so Q' would contain 3 points. Let $Q' = \{A, B\}$. There are 3 points on the line at infinity and not on AB: any two of these together with A, B give a completion of Q'.

§ 6.1

2. Take $a = (1, 0, \ldots, 0)$, $b = (0, 1, 0, \ldots, 0)$. Then D.6 holds for a, b if, and only if, $2^{2/p} = 2$, i.e., $p = 2$.

4. Write $\|a\|$ for $d(a, 0)$. We are asked to prove that

$$\sigma(a + b, c) - \sigma(a, c) - \sigma(b, c) = 0,$$

i.e.,

$$\|a + b + c\|^2 - \|a + b\|^2 - \|a + c\|^2 - \|b + c\|^2 + \|a\|^2 + \|b\|^2 + \|c\|^2 = 0.$$

But

$$\|a + b + c\|^2 + \|a + b - c\|^2 = 2\|a + b\|^2 + 2\|c\|^2,$$
$$\|a + b + c\|^2 + \|a - b + c\|^2 = 2\|a + c\|^2 + 2\|b\|^2,$$
$$2\|a\|^2 + 2\|b - c\|^2 = \|a + b - c\|^2 + \|a - b + c\|^2,$$
$$4\|b\|^2 + 4\|c\|^2 = 2\|b + c\|^2 + 2\|b - c\|^2.$$

Adding these equations gives the result.

5. We write $\|a\|$ for $d(a, 0)$. Then

$$\sigma(xa, b) = \|xa + b\|^2 - \|xa\|^2 - \|b\|^2.$$

It is clearly sufficient to show that $x \to \|xa + b\|$ is a continuous function (for any given a, b in V). The triangle inequality shows that

$$\big| \, \|xa + b\| - \|ya + b\| \, \big| \leq \|xa - ya\|$$
$$= |x - y| \, \|a\|, \quad \text{by D.5,}$$

and so $\|xa + b\| \to \|ya + b\|$ as $x \to y$.

Take any x in $\mathbf{R}$ and any sequence (r_i) of rational numbers so that $\lim_{i \to \infty} r_i = x$. Then

$$\sigma(xa, b) = \lim_{i \to \infty} \sigma(r_i a, b)$$
$$= \lim_{i \to \infty} r_i \sigma(a, b) \quad \text{(exercise 4)}$$
$$= x \, \sigma(a, b).$$

7. Let $u = b - c$, $v = a - b$. Then

$$\|u + v\|^2 = (\|u\| + \|v\|)^2 = \|u\|^2 + \|v\|^2 + 2\|u\| \, \|v\|.$$

Hence $\sigma(u, v) = \|u\| \, \|v\|$ and, by Schwarz's Inequality (Lemma 1), u, v are linearly dependent. Thus a, b, c are linearly dependent.

Let $S = a + M$ and $S' = a' + M'$, where we choose $a' = a\varphi$. Then φ induces a one-one mapping ψ of M onto M':

$$v \to (a + v)\varphi - a'.$$

Thus $0\psi = 0$ and ψ preserves distance: if $v_1, v_2 \in M$,

$$d(v_1, v_2) = d(a + v_1, a + v_2)$$
$$= d'((a + v_1)\varphi, (a + v_2)\varphi)$$
$$= d'(v_1\psi, v_2\psi).$$

By the first part (and its converse) ψ maps lines to lines. Hence for any $v \neq 0$ in M and any x in $\mathbf{R}$, $(xv)\psi$ lies on the line $[v\psi]$: say $(xv)\psi = x'(v\psi)$. Since $d(v, 0) = d'(v\psi, 0)$ and $d(xv, 0) = d'(x'(v\psi), 0)$, $|x| = |x'|$. Also $d(xv, v) = d'(x'(v\psi), v\psi)$ and so $|x - 1| = |x' - 1|$. Thus $x = x'$ and $(xv)\psi = x(v\psi)$. Further, if u, v are arbitrary vectors in M, let w be the mid-point of u, v, i.e., $w = \frac{1}{2}(u + v)$. Then $w\psi$ is the mid-point of $u\psi, v\psi$. Hence $(u + v)\psi = (2w)\psi = 2(w\psi) = u\psi + v\psi$. Thus ψ is a linear mapping.

§ 6.2

4. Let $(a_1, \ldots, a_n)$ be a cartesian basis of (H, σ). If $Q(\tau) \cap A$ is a sphere, then the equation of $Q(\tau)$ (with respect to a projective coordinate system determined by $(a_0, \ldots, a_n)$) is $X_1^2 + \cdots + X_n^2 = k^2 X_0^2$ (cf. p. 117). If P is the point $[(0, x_1, \ldots, x_n)]$, then $[(y_0, y_1, \ldots, y_n)] \subset P^{\perp(\tau)}$ if, and only if, $x_1 y_1 + \cdots + x_n y_n = 0$, i.e., if, and only if, $[(0, y_1, \ldots, y_n)] \subset P^{\perp(\sigma)}$.

Conversely, we assume that $\perp(\tau_H) = \perp(\sigma)$ and so $H^{\perp(\tau)} \cap H = 0$. Thus $V = H \oplus H^{\perp(\tau)}$ and $Q(\tau)$ has equation $X_1^2 + \cdots + X_n^2 = sX_0^2$. Since $Q(\tau)$ has more than one point we must have $s > 0$ and so $Q(\tau)$ is a sphere in A (as defined in exercise 3).

5. There is a unique translation taking B_1 to A_1. Let it take B_2 to A_3. Then $d(B_1, B_2) = d(A_1, A_3)$. Take the sphere with centre A_1 and containing A_2. This meets $A_1 A_3$ in two points. Let A_4 be one of them. If A_0 is the point at infinity on $A_1 A_3$,

$$d(A_1, A_2)/d(B_1, B_2) = d(A_1, A_4)/d(A_1, A_3)$$
$$= |\text{cr}(A_0, A_1; A_3, A_4)| \quad \text{by exercise 2}.$$

§ 6.3

3. Let c be a common eigenvector of f and g and let M be a subspace for which $V = M \oplus [c]$. If mf is expressed as $mf' + xc$ where m, $mf' \in M$ and $x \in F$, it is easy to check that f' is a linear mapping of M into M and $(fg)' = f'g'$. Now use induction on dim V.

§ 6.5

3. The given equation is equivalent to $f = \tau g^{-1}$. Let a_1 be an eigenvector of f and without loss of generality assume $\sigma(a_1, a_1) = 1$. Now $\sigma(a_1, b) = 0$ implies $\sigma(a_1 f, b) = 0$ and so $\tau(a_1, b) = 0$. We have $W = [a_1] \oplus [a_1]^{\perp(\sigma)}$ and $[a_1]^{\perp(\sigma)} \subset [a_1]^{\perp(\tau)}$. Now use induction on dim W.

4. As for exercise 3 of § 6.3 but with $M = [c]^{\perp(\sigma)}$ and $\sigma(c, c) = 1$.

5. The given equation is equivalent to $b(af^*g) = bf(ag)$ for all a, b, i.e., $af^* = (f(ag))g^{-1}$ for all a. Since g is semi-linear with respect to complex conjugation this defines a linear mapping f^*. (In the language of exercise 7, § 4.6, p. 83, $f(ag) = a(gf^t)$ and so $f^* = gf^tg^{-1}$.)
   (i)
   $$\sigma(af^{**}, b) = \sigma(a, bf^*)$$
   $$= \overline{\sigma(bf^*, a)}$$
   $$= \overline{\sigma(b, af)}$$
   $$= \sigma(af, b)$$
   for all a, b; hence $f^{**} = f$.
   (ii), (iii) and (iv) are proved similarly; (v) is a consequence of (ii) and (iv).

6. (i)
   $$\sigma(wf^*, wf^*) = \sigma(w, wf^*f) = \sigma(w, wff^*) = \overline{\sigma(wff^*, w)}$$
   $$= \overline{\sigma(wf, wf)} = \sigma(wf, wf).$$
   (ii) If $w(f - x1) = 0$, then $w(f - x1)^* = 0$ by (i). Hence the result by (v) of exercise 5.
   (iii) If $b \in [w]^\perp$, $\sigma(wf^*, b) = 0$ because $[w]f^* \subset [w]$ by (ii). So $\sigma(w, bf) = 0$, as required.

§ 7.3

2. If z is non-zero, non-invertible and not irreducible, $z = xy_1$ where x, y_1 are non-zero and non-invertible. Hence Ry_1 is strictly between Rz and R. If y_1 is not irreducible, we obtain y_2 so that Ry_2 is strictly between Ry_1 and R. By exercise 1, this process must terminate and hence z has an irreducible factor.
   Let $z = z_1x_1$, where z_1 is irreducible. If x_1 is non-invertible, $x_1 = z_2x_2$ where z_2 is irreducible. Again we obtain a strictly ascending chain of ideals $Rz \subset Rx_1 \subset Rx_2 \subset \cdots$. This can only terminate if there exists r such that x_r is irreducible. Then $z = z_1 \ldots z_{r-1}x_r$, as required.

4. If p is irreducible and p divides ab where $ab \neq 0$ and p does not divide a, then p divides b. For if $ab = cp$ and $up + va = 1$, then $b = upb + vab = (ub + vc)p$.

Applying this result to $p_1 \cdots p_r = q_1 \cdots q_s$, we see that p_1 divides some q_j, say q_1. Since p_1, q_1 are both irreducible, $p_1 = e_1 q_1$ for some invertible element e_1. Cancelling q_1 we obtain $e_1 p_2 \cdots p_r = q_2 \cdots q_s$ and the result now follows by an induction on $\min(r, s)$.

7. In part (b) (i) of the proof of Theorem 1 we argue thus: If

$$[d] = [m_j] + [m_k] \quad \text{and} \quad d = um_j + vm_k,$$

put

$$b_j{}' = vb_j - ub_k,$$

$$b_k{}' = \frac{m_j}{d} b_j + \frac{m_k}{d} b_k.$$

Then $[b_j, b_k] = [b_j{}', b_k{}']$ and $m_j b_j + m_k b_k = d b_k{}'$. Hence d has length $\geq$ length m_k. Thus $xd = m_k$, where x is invertible (exercise 5) and so m_k divides m_j.

Part (b) (ii): Let d be a greatest common divisor of m_k, n_k and suppose $d = um_k + vn_k$. Then u times the first relation plus v times the second yields

$$\sum_{i=1}^{k-1} (um_i + vn_i) b_i + db_k = 0$$

and so (again by exercise 5) $m_k = xd$, where x is invertible. Thus m_k divides n_k.

11. If m is the positive generator of the smallest invariant factor of the multiplicative group F^* viewed as a $\mathbf{Z}$-module, then $x^m = 1$ for all x in F^* and there exists an element x_0 whose annihilator is $\mathbf{Z}m$, i.e., whose order is m.

Thus all elements of F are roots of $X^{m+1} - X$. But this polynomial has at most $m + 1$ roots and F has at least $m + 1$ elements. Therefore $q = m + 1$.

12. (i) E is closed under addition, subtraction, multiplication and inversion by non-zero elements and E contains $1, 2, \ldots, p-1$.

(ii) Let $k = lm$. Then $p^m - 1$ divides $p^k - 1$ and hence (by the same argument) $X^{p^m - 1} - 1$ divides $X^{p^k - 1} - 1$. Thus $X^{p^m} - X$ divides $X^{p^k} - X$. By exercise 11, all the elements of F are roots of $X^{p^k} - X$ and hence the subset E consisting of the roots in F of $X^{p^m} - X$ has exactly p^m elements. This is a subfield by (i).

If E' is a field with p^m elements contained in F, then all elements of E' are roots of $X^{p^m} - X$ and so $E' = E$.

(iii) By exercise 10, F is a k-dimensional vector space over $\mathbf{F}_p$. But also E is an m-dimensional vector space over $\mathbf{F}_p$ and F is a vector space over E, necessarily finite dimensional, say of dimension l. Hence F is an ml-dimensional vector space over $\mathbf{F}_p$ and thus $ml = k$.

§ 7.4

2. Let $M = E/S$ and $c_i = v_i + S$. Every elementary move on $(c_1, \ldots, c_n)$ is obtained by an elementary move on $(v_1, \ldots, v_n)$. Hence exercise 1 yields the result.

§ 7.5

3. Apply elementary moves (over the smaller field F) to the rows and columns of the matrix $X1 - A$ to obtain an equivalent matrix of the form $\text{diag}(d_1, \ldots, d_t)$ (as in Proposition 8). The invariant factors are then $[d_{s+1}] \supset \cdots \supset [d_t]$ relative to F or to any field containing F. The minimum polynomial is d_t (to within a non-zero constant multiple).

The elementary divisors of

$$\begin{pmatrix} 0 & 1 \\ 2 & 0 \end{pmatrix}$$

are (i) $X^2 - 2$ relative to $\mathbf{Q}$ and (ii) $X - \sqrt{2}$, $X + \sqrt{2}$ relative to $\mathbf{R}$.

§ 7.6

4. Complexify and note that a real invariant line always contains a pair of conjugate complex invariant points.

List of Symbols

Sets

Z	integers	1
Q	rational numbers	1
R	real numbers	1
C	complex numbers	1
$\mathbf{F}_p$	integers modulo p	1
$x \in S$	x is an element of S	1
$S \subset T$	S is contained in T	1
$S = T$	S equals T	2
$\{x, y, \ldots\}$	the set consisting of $x, y, \ldots$	2
$\neq$	is not equal to	2
$\not\subset$	is not contained in	2
$\notin$	is not an element of	2
$f: S \to T$	mapping f of S into T	2
Sf	the image of S under f	2
f^{-1}	the inverse of f	3 (ex. 1)

Index Notation

$(f_s)_{s \in S}$	family	2
$\bigcap(M_i : i \in I), M_1 \cap \cdots \cap M_n$	intersection	3
$\bigcup(M_i : i \in I), M_1 \cup \cdots \cup M_n$	union	3
$+(M_i : i \in I), M_1 + \cdots + M_n$	sum	7
$\bigoplus(M_i : i \in I), M_1 \oplus \cdots \oplus M_n$	direct sum	7
$\mathsf{J}(S_i : i \in I), S_1 \mathsf{J} S_2$	join	16

Vector Spaces

0_V	zero element of V	5
F^n	n-tuples	5
$F[X]$	polynomials in X	6
$[S]$	subspace spanned by S	7
$\dim_F V$	dimension of V over F	10
$\{e_1, \ldots, e_n\}$	standard basis of F^n	13

188

$U_{(c)}$	complexification of real space U	14
t_a	translation $v \rightarrow v + a$	42
Ker f	kernel of f	66
V/K	space of cosets of K	67
$\mathscr{L}_F(V, V')$	linear mappings of V into V'	68
$F^{m \times n}$	$m \times n$ matrices	72
$\{E_{ij}\}$	standard basis of $F^{m \times n}$	74 (ex. 2)
V^*	dual space of V	80
$(a^1, \ldots, a^n)$	dual basis of $(a_1, \ldots, a_n)$	80
M°	annihilator of M	81
ι	canonical isomorphism $V \rightarrow V^{**}$	81
f^t	transpose of f	83 (ex. 7)
$\mathscr{B}(V)$	bilinear forms on V	89
$\underline{\sigma}, \tilde{\sigma}$	linear mappings induced by the bilinear form σ	91
$M^{\perp(\sigma)}, M^{\top(\sigma)}$	subspaces orthogonal to M	93
σ_M	restriction of σ to subspace M	95
$\mathscr{B}^+(V)$	symmetric bilinear forms on V	101 (ex. 4)
$\mathscr{B}^-(V)$	skew-symmetric bilinear forms on V	101 (ex. 4)
$\mathscr{Q}(V)$	quadratic forms on V	102 (ex. 9)
$E_x(f)$	eigenspace of f corresponding to eigenvalue x	135

Geometries

$\mathscr{A}(S)$	affine geometry on coset S	16
$\mathscr{P}(V)$	projective geometry on vector space V	29
$(\wedge, \varphi)$ or A	generalized affine geometry	35
(P, ψ) or P	generalized projective geometry	35
$\mathscr{A}(f)$	mapping of affine geometries induced by $f: S \rightarrow S'$	42
$\mathscr{A}(t_{-a} g\, t_{a'})$	affinity	42
$\mathscr{P}(f)$	mapping of projective geometries induced by $f: V \rightarrow V'$	45
$\mathfrak{P}^*$	dual proposition	84
$\mathfrak{C}^*$	dual configuration	84
$\mathscr{P}^*(V)$	dual geometry	86
$Q(\sigma)$	quadric	111
$C(Q)$	representative cone of projective quadric Q	115
$\|a\|$	norm of a	125
$d(a, b)$	distance from a to b	125
$\mathscr{A}(S, d)$	euclidean geometry on coset S with distance d	127
$\widehat{BAC}$	angle	128
(d)	class of distances similar to d	131

$\mathscr{A}(S, D)$	similarity euclidean geometry on coset S with distance class D	131
(A, D_φ) or $(A, \perp)$	generalized similarity euclidean geometry	131, 132
$\mathrm{cr}(A_0, A_1; A_2, A_3)$	cross-ratio	133 (ex. 1)

Groups

Aut X	automorphisms on X; $X = V, F$, A or P	59
Aut V	general linear group	59
$S(V)$	semi-linear automorphisms of V	59
Tr V	translations on V	59
Af A	affine group on A	59
Tr A	translation in A	59
Pr P	projective (collineation) group on P	59
$(\mathrm{Aut}\ \mathscr{P}(V))_H$	automorphisms of $\mathscr{P}(V)$ leaving H invariant	61
$(\mathrm{Pr}\ \mathscr{P}(V))_H$	collineations of $\mathscr{P}(V)$ leaving H invariant	61
Aut(V, σ)	orthogonal group on euclidean space (V, σ)	144
Aut(V, D)	automorphisms of V preserving D	144
$(\mathrm{Pr}\ \mathscr{P}(V))_{H,\perp}$	collineations of $\mathscr{P}(V)$ which restrict to collineations of $\mathscr{P}(H)$ preserving $\perp$	145
Aut(W, σ)	unitary group on Hilbert space (W, σ)	146

Matrices

$(f; (v_i), (v_j'))$	matrix of f with respect to ordered bases (v_i), (v_j')	72
0	zero matrix	72
I	identity matrix	73
A^{-1}	inverse of A	73
δ_{ij}	Kronecker delta	73
E_{ij}	matrix with (i, j)-term 1 and others 0	74 (ex. 2)
A^t	transpose of A	77
diag $(a_{11}, \ldots, a_{nn})$	diagonal matrix	104
$A\zeta$	matrix $(a_{ij}\zeta)$	124

Modules

V_f	$F[X]$-module defined by $f: V \to V$	151
$m(X)$	minimum polynomial	151
$m_a(X)$	minimum polynomial at a	151
$\mathfrak{a}(S)$	annihilator (ideal) of S	151
$[S]_f$	submodule of V_f generated by S	153
M/K	module of cosets of K	154

Bibliography

[1] E. Artin, *Geometric algebra* (Interscience).

[2] G. Bachmann and L. Narici, *Functional analysis* (Academic Press).

[3] R. P. Burn, *Deductive transformation geometry* (Cambridge).

[4] H. S. M. Coxeter, *Introduction to geometry* (Wiley).

[5] J. Dieudonné, *Sur les groupes classiques* (Hermann).

[6] D. R. Hughes and E. C. Piper, *Projective planes* (Springer).

[7] I. Kaplansky, *Linear algebra and geometry* (Allyn and Bacon).

[8] F. Riesz and B. Sz.-Nagy, *Functional analysis* (Blackie).

[9] A. Tuller, *A modern introduction to geometries* (van Nostrand).

Index

Affine geometry (*see* Geometry)
Affinity, 42ff
  degenerate, 70ff
Angle, 126, 128
Annihilator, 81
  ideal, 151
  mapping, 81, 94
Anti-automorphism, 120, 122
Anti-isomorphism, 84
Associative law
  for mapping multiplication, 3 (ex. 3)
  for matrix multiplication, 73
Automorphism
  of euclidean geometry, 140ff
  of field, 56, 58 (incl. exx. 1, 2), 65, 123
  of Hilbert space, 146
  of R-module, 154
  of sesquilinear space, 124
  of similarity euclidean geometry, 144
  of vector space, 66
  groups for V, A, P, F, 59
Axis, of central collineation, 48 (exx. 3–5)

Basis (base), 9ff
  cartesian, 127, 142, 146
  dual, 80
  standard, of F^n, 12 (ex. 9), 13
  standard, of $F^{m \times n}$, 74 (ex. 2)
  standard, of $F[X]^n$, 168
  of torsion-free $F[X]$-module, 163
Bilinear form, 89ff
  alternating, 99
  degenerate, 91, 92 (ex. 1)
  negative definite, 106
  orthosymmetric, 97, 101 (ex. 3)
  positive definite, 106
  skew-symmetric, 99, 101 (exx. 1, 4)
  symmetric, 97, 101 (exx. 1, 4, 8, 9)
Bilinear polynomial, 90

Canonical matrix
  Jordan (classical), 167
  rational, 166
Cartesian basis, 127, 142, 146
Center
  of affine polarity, 111 (ex. 5)
  of affine quadric, 119
  of (central) collineation, 48 (exx. 3–6), 54 (ex. 1), 63, 83 (ex. 10)
  of perspective of two triangles, 23
  of perspectivity, 48 (ex. 6), 65
  of sphere, 133 (ex. 3)
Central collineation, 48 (exx. 3–6), 54 (ex. 1), 62ff, 83 (ex. 10)
  center of (*see* Center)
  hyperplane of, 63
Centroid, 48 (ex. 2)
Characteristic, 2
  field of, 40, 99ff
  (Fano's Theorem), 40
  [*Note rubrics at the beginning of* § 5.6 (p. 111) *and* § 5.7 (p. 115).]
Classical (Jordan) canonical matrix, 167
Classification of collineations, 170ff
Coefficient space, 50
Collineation, 45, 170ff
  central (*see* Central)
Completeness of Hilbert space, 147
Completion of affine quadric (to projective quadric), 117, 118, 120 (ex. 8)
Complexification
  of real bilinear form, 124, 146
  of real vector space, 14, 58 (ex. 4), 112, 134 (ex. 6), 136, 140
Cone, representative, 115
Congruence
  of bilinear forms, 89, 90, 103ff
  of matrices, 86, 90, 103ff
Conic, 111, 114 (exx. 4–10), 120 (exx. 1, 2)
  section, 116

Coordinate row
  of an affine point, 44
  of a vector, 13
  (homogeneous) of a projective
    point, 47
Coordinate system
  for an affine geometry, 44
  for a euclidean geometry, 129
  (homogeneous) for a projective
    geometry, 46
  for a vector space, 13
Correlation, 85, 110, 120
  degenerate, 110
Coset (translated subspace), 15ff
  dimension of, 16
  parallel, 18
  space of, 67 154
  subspace belonging to, 16
Cross-ratio, 132, 133 (exx. 1, 2, 5,
    6)

Degenerate
  affine polarity, 111 (ex. 6)
  affine quadric, 119
  affinity, 70
  bilinear form, 91, 92 (ex. 1)
  correlation, 110
  null-polarity, 110
  polarity, 110
  projective quadric, 112
  projectivity, 70
  subspace (with respect to a bilinear
    form), 92 (ex. 1), 95
Desargue's Theorem
  (affine), 23
  (projective), 32 (ex. 6), 38, 41 (ex.
    3), 48 (ex. 5), 51 (ex. 2), 86
    (ex. 1)
  converse of, 38
  related to a division ring, 39, 40,
    48 (ex. 5)
Determinant, 144
Dilatation, 44 (ex. 3), 49 (ex. 9), 54
    (ex. 1), 63, 75 (ex. 11)
Dimension
  of an affine geometry, 16
  of a coset, 16
  of a projective geometry, 30
  of a vector space, 10ff
  finite, 9ff
    (Note the convention at the begin-
      ning of Chapter II, p. 15.)
  projective, 30
Direct sum
  of submodules, 155ff
  of subspaces, 7, 8 (exx. 7, 8), 11, 12
    (exx. 11–13), 95, 96, 102ff, 155ff

Distance
  on a coset, 127
  on a euclidean space, 125
  on a Hilbert space, 146
  on a vector space, 125
  similar, 131
Division ring, 39, 49 (exx. 10, 11), 75
    (ex. 7)
Dual
  basis, 80
  configuration, 84
  frame of reference, 87, 88 (ex. 1)
  geometry, 86ff
  proposition, 84
  self-, 171, 172
  space, 80
Duality, 85
  principle of, 84

Eigenspace, 71 (ex. 4), 135, 142, 143
Eigenvalue, 71 (ex. 4), 135, 137, 141ff
Eigenvector, 71 (ex. 4), 135ff, 140
    (ex. 3), 141ff
Elementary divisors
  of a finitely generated $F[X]$-
    module, 159, 160
  of a linear mapping, 166
  of a matrix, 168, 169 (ex. 3)
Elementary move, 164, 165 (ex. 1, 2)
Elementary transformation, 70 (ex.
    10)
Embedding Theorem, 32, 52 (ex. 6),
    61, 116, 130
Equation
  of affine quadric, 117ff
  of euclidean quadric, 138ff
  of projective quadric, 112ff
  linear, 49ff, 79ff, 82
Equivalence of matrices, 77, 165
Exchange Lemma, 9
Exponent of commutative group, 153
    (ex. 4)

Family, 2
Fano's Theorem, 40
Field, 4, 5
  of characteristic 2, 40 99ff (conven-
    tions on pp. 111, 115)
  of gaussian numbers, 14 (ex. 1)
  of quotients, 160
  finite (see Finite field)
  ground, 12
Finite dimensional vector space, 9ff
    (Note the convention at beginning
      of Chapter II, p. 15.)
Finite field, 109 (exx. 8–11), 162
    (exx. 10–12), 174 (ex. 5)

Finite geometry
 (affine), 20 (ex. 1), 29, 38 (ex. 3)
 (projective), 32 (exx. 4, 5), 40, 41,
 119, 174 (ex. 5)
Form
 bilinear (see Bilinear form)
 hermitian, 122ff
 linear, 80ff, 101 (ex. 7)
 quadratic, 100ff
 sesquilinear, 121ff
Frame of reference
 for an affine geometry, 43
 for a euclidian geometry, 129
 for a projective geometry, 46
 for $\mathscr{P}(V)$ determined by a basis of
  V, 46
 standard, for $\mathscr{A}(F^n)$, 43
 standard, for $\mathscr{P}(F^{n+1})$, 47
Function (see Mapping)

Gaussian number field, 14 (ex. 1)
Generator of a projective quadric, 114
 (exx. 11–14)
Geometry
 affine, 16ff (generalized), 35
 affine, in $\mathscr{P}(V)$ determined by H,
  36, 61, 130
 "affine, of dimension n over F," 22
 euclidean 127ff
 finite (see Finite geometry)
 projective, 29ff (generalized), 35
 "projective, of dimension n over
  F," 32
 similarity euclidean, 131ff
 similarity euclidean, determined in
  $\mathscr{P}(V)$ by H and D, 132
Greatest common divisor, 161 (exx.
 3, 6)
Ground field, 12ff
Group, 4
 abelian (commutative), 4, 150, 160,
  163 (ex. 13)
 affine, 59, 75
 automorphism, 59
 collineation (projective), 59
 commutative (see Group, abelian)
 general linear, 59
 homomorphism of, 59
 isomorphism of, 59
 orthogonal, 144
 permutation, 4, 62 (ex. 4)
 projective (see Group, collineation)
 rotation, 4, 144
 unitary, 146

Harmonic
 conjugate, (affine) 27, (projective) 40

constructive, (affine) 27, (projective) 40, 52 (ex. 5)
range, (affine) 27, (projective), 40,
 113 (ex. 2), 133 (ex. 1)
Hermitian
 form, 122ff, 124 (exx. 2, 3)
 mapping, 147 (ex. 6)
 matrix, 124 (ex. 2)
Hilbert space, 146ff
Homogeneous
 (projective) coordinate row, 47
 linear equations, 50
 vector, 22ff, 30
Homomorphism
 of groups, 59
 of R-modules, 154
 of vector spaces, 66
Hyperplane
 in affine geometry, 16
 of a central collineation, 63ff
 at infinity, 36
 in projective geometry, 30

Ideal, 152
 principal, 152
 annihilator, 151
Identity
 element of group, 4
 element of ring, 149
 mapping, 2, 3 (ex. 1)
 matrix, 73
Image, 2
Incidence propositions, (affine) 19,
 (projective) 31
Indecomposable submodule, 159
Index notation, 2
Infinite dimensional vector space, 9,
 11 (exx. 3, 5, 7, 8), 68, 69 (exx.
 1, 2, 4, 9), 82 (exx. 1, 3, 5, 6), 92
 (exx. 1, 2, 3), 130 (ex. 4), 147
Integral domain, 155 (exx. 1, 3), 160
Intersection
 of cosets, 16
 of sets, 3
 of subspaces, 7
Invariant configuration, 170ff
Invariant factors
 of linear mapping, 166
 of matrix, 168, 169 (ex. 3)
 of module, 156, 160, 165
Invariant subspace, 153
Inverse
 of a group element, 4
 of a mapping, 3 (ex. 1).
 of a matrix, 73, 74
Isomorphism
 of affine geometries, 20ff, 52ff

Isomorphism—*cont.*
 of bilinear spaces, 89
 of euclidean geometries, 128, 130
  (ex. 7)
 of fields, 56
 of groups, 59
 of modules, 154
 of projective geometries, 31, 52ff
 of similarity euclidean geometries,
  131
 of vector spaces, 12
 anti-isomorphism, 84
 linear isomorphism, 66, 91
 semi-linear isomorphism, 56

Join
 of cosets, 16
 (sum) of subspaces, 7
Jordan (classical) canonical matrix,
 167

Kernel
 of group homomorphism, 60, 62
  (ex. 4)
 of linear mapping, 66
 of module homomorphism, 154
Kronecker delta, 69 (ex. 6), 73

Length
 (norm) in euclidean space, 125
 (norm) in Hilbert space, 146
 of element in a principal ideal
  domain, 161 (exx. 5, 7)
Line
 in affine geometry, 16
 at infinity, 36, 37
 projective, 30
Linear algebra, 68, 136
Linear automorphism, 66
Linear combination, 6 (ex. 5)
Linear complex, 110 (ex. 4)
Linear dependence, 8ff
Linear equations, 49ff, 79ff, 82
Linear form, 80ff
Linear independence
 over F, 9
 over $F[X]$, 158
Linear isomorphism, 12, 66, 91
Linear mapping, 66
 commuting, 140 (ex. 3), 147 (ex. 4)
 σ-symmetric, 135, 140 (ex. 1)
 similar, 154
 trace of, 75 (exx. 8, 9), 83 (ex. 9)
 transpose of, 83 (exx. 7, 8), 93 (ex.
  4), 169
 triangulable, 140 (ex. 3), 147 (ex. 4)

Mapping (function), 2
 annihilator, 81, 94

hermitian, 147 (ex. 6)
inverse, 3 (ex. 1)
linear (*see* Linear mapping)
normal, 147 (ex. 6)
one—one, 2
product of, 2
semi-linear, 120
unitary, 147 (ex. 6)
Matrix, 72ff
 addition of, 72
 alternating, 103
 of bilinear form, 90
 congruence of, 86, 103ff
 diagonal, 104
 equivalence of, 77, 165
 hermitian, 124 (ex. 2)
 identity, 73
 inverse of, 73, 74
 invertible (non-singular), 74
 Jordan canonical, 167
 of linear mapping, 72
 multiplication of, 72
 nilpotent, 74
 non-singular (invertible), 74
 orthogonal, 79 (exx. 2, 3), 143
 of quadratic form, 102
 rational canonical, 166
 similarity of, 79, 166ff
 skew-symmetric, 79 (exx. 1, 3), 101
  (ex. 1)
 symmetric, 77, 101 (ex. 1)
 transposed, 77
 triangular, 108, 109 (ex. 7), 140 (ex.
  3), 147 (ex. 4)
 unitary, 146
 zero, 72
Metric, 125
Mid-point, 29
Minimum polynomial
 of linear mapping, 151
 of matrix, 168, 169 (ex. 3)
Minkowski's inequality, 129 (ex. 2)
Module, 150
 cyclic, 153
 finitely generated, 153
 torsion, 152, 155 (exx. 1, 2), 165
  (ex. 2)
 torsion-free, 152, 163, 165 (exx. 2,
  3)

Nilpotent matrix, 74 (ex. 4)
Non-Desarguesian projective plane,
 39, 48 (ex. 5)
Norm (length)
 in euclidean space, 125
 in Hilbert space, 146
Normal mapping, 147 (ex. 6)

Normal to a plane, 134 (ex. 7)
n-tuple, 2
  ordered, 2
Null-polarity, 110

Order
  of (commutative) group, 153 (ex. 4)
  of general linear group, 62 (ex. 3)
Orientation, 144
Origin of affine frame of reference, 43
Orthogonality, 93ff, 126, 137, 146 (anticipated, 51)
Orthosymmetry, 97, 101, (ex. 3), 122

Pappus' Theorem, (affine) 25, (projective) 39, 40, 51 (ex. 3), 88 (ex. 5)
Paraboloid, 119
Parallel, 18
Parallelogram, 37, 64
  law, 125
Pencil
  of lines, 86 (ex. 3), 87
  of planes, 86 (ex. 3), 87
Permutation, 4
Perpendicular, 133, 134 (ex. 7)
Perspective
  axis of, 38
  center of, 23
  triangles in, from a point, 23
Perspectivity, 48 (exx. 6–8), 49 (ex. 11), 65
Plane
  in affine geometry, 16
  projective, 30
Point
  in affine geometry, 16
  at infinity, 36
  projective, 30
Polarity, 110, 113 (exx. 2, 3, 4, 7, 8, 16, 17)
  affine, 111 (exx. 5, 6), 120 (ex. 2)
  definite, 132
Polynomials, vector space of, 6
Positive definite
  hermitian form, 123
  symmetric bilinear form, 106
Principal ideal, 152
  domain 160, 161 (exx. 1–8)
Projection, 70 (ex. 9), 71 (ex. 3)
Projective geometry (see Geometry)
Projective space, 30
Projectivity, 45ff, 88
  degenerate, 70ff
Pythagoras' Theorem, 100

Quadrangle, 40, 84
Quadratic
  form, 100, 101 (ex. 5–9)
  polynomial, 100ff, 108 (ex. 5), 111ff, 117
Quadric
  affine, 116ff
  euclidean, 134ff
  projective, 111ff
Quadrilateral, 84
Quaternion algebra (over $\mathbf{R}$), 75 (ex. 7), 101 (ex. 6)

Radius of sphere, 133 (ex. 3)
Range of points, 86 (ex. 3), 87
Rank
  of affine quadric, 119
  of bilinear form, 91
  of linear mapping, 76ff, 94
  of matrix (including row and column), 77
  of projective quadric, 112
  of sesquilinear form, 122
  of torsion-free $F[X]$-module, 163
Rational canonical matrix, 166
Reflexion, 143, 145 (ex. 4)
Relativity, 107
Representative cone, 115
Ring, 149ff
Rotation, 143, 145 (exx. 1, 2, 3)
  group, 4, 144

Scalar, 5
Schwarz's inequality, 126
Segre symbol, 170
Self-polar simplex, 115 (ex. 15)
Semi-linear
  isomorphism, 56
  mapping, 120
Sequence, 2
  ordered, 2
Sesquilinear form, 121ff
  orthosymmetric, 122
Sets, 1
  equality of, 2
Signature, 106, 108 (ex. 4)
Similarity
  of distances, 131
  of linear mappings, 154
  of matrices, 79, 166ff
Simplex of reference, 46
Skew
  elements in projective geometry, 30
  lines in affine geometry, 20
Space
  alternating (symplectic), 99, 102ff
  bilinear, 89

Space—*cont.*
  coefficient, 50
  of cosets, 67, 154
  dual, 80
  eigen, 71 (ex. 4), 135
  euclidean, 107, 125ff
  Hilbert, 146ff
  metric, 125
  normed, 125
  orthogonal (symmetric), 98, 105ff
  projective, 30
  skew-symmetric, 99
  symmetric (orthogonal), 98, 105ff
  symplectic (alternating), 99, 102ff
  vector, 5ff
Space time, 107
Sphere, 133 (exx. 3, 4)
Split product of groups, 60, 144
Standard basis
  of F^n, 12 (ex. 9), 13
  of $F^{m \times n}$, 74 (ex. 2)
  of $F[X]^n$, 168
Subgeometry
  of affine geometry, 16
  of euclidean geometry, 127
  of projective geometry, 30
  of similarity euclidean geometry,
      131
Subgroup, 4, 59ff, 144ff
  Sylow, 163 (ex. 13)
Submodule, 153ff
Subring, 149
Subspace, 6ff
  belonging to a coset, 16
  direct sum of, 7
  invariant, 153
  spanned (generated) by a set, 7
  sum (join) of, 7
  translated (*see* Coset)
Sylvester's Law of Inertia, 106

Tangent to a projective quadric, 114
  (exx. 9–11)
Torsion element, 152, 155 (ex. 3)
Torsion-free $F[X]$-module, 163, 165
  (exx. 2, 3)
Torsion module (*see* Module)
Trace mapping, 75 (exx. 8, 9), 83 (ex.
  9)
Translation
  in affine geometry, 42, 49 (ex. 9),
      54 (ex. 1), 63, 70 (ex. 11)
  group (in A and in V), 59
  in vector space, 42
Transpose
  of linear mapping (*see* Linear
      mapping)
  of matrix, 77
Transvection, 70 (ex. 11)
Triangle, 23
  diagonal point, 40
  inequality, 125

Union of sets, 3
Unique factorization domain, 160,
  161 (ex. 4)
Unit point of a projective frame of
  reference, 46

Vector, 5
  eigen (*see* Eigenvector)
  homogeneous, 22ff, 30
Vector space, 5ff
  dimension of, 10ff
  finite dimensional, 9
  infinite dimensional (*see* Infinite
      dimensional vector space)

Wedderburn's Theorem, 39

Zorn's Lemma, 178 (solution of ex. 6
  of § 4.6)